Mechanisms and Machine Science

Volume 19

Series Editor

Marco Ceccarelli, Cassino, Italy

For further volumes:
http://www.springer.com/series/8779

Juan Carlos García-Prada · Cristina Castejón
Editors

New Trends in Educational Activity in the Field of Mechanism and Machine Theory

Springer

Editors
Juan Carlos García-Prada
Cristina Castejón
Department of Mechanical Engineering
Universidad Carlos III de Madrid
Leganés
Spain

ISSN 2211-0984 ISSN 2211-0992 (electronic)
ISBN 978-3-319-37570-0 ISBN 978-3-319-01836-2 (eBook)
DOI 10.1007/978-3-319-01836-2
Springer Cham Heidelberg New York Dordrecht London

Printed on acid-free paper

Springer is part of Springer Science+Business Media (www.springer.com)

Preface

Proceedings of ISEMMS

Education is of primary interest for any academic community. This symposium is the first of a series that is organized by Permanent Commission (PC) for Education of IFToMM, the International Federation for the Promotion of Mechanism and Machine Science (MMS). IFToMM has addressed attention to Education of MMS since its beginning in 1969 with discussions and plans in several frames. After many years and several attempts, the IFToMM Executive Council has supported the organization of ISEMMS, IFToMM Symposium on Education of MMS since 2010 thanks to the efforts by Prof. Juan-Carlos García-Prada in his position of PC chair. The team at the Carlos III University in Leganés, Spain, as coordinated locally by Prof. Cristina Castejón did a great job attracting several interesting contributions in the many aspects of MMS Education. Authors from all around the world have presented their views and information about current situations and possible future perspectives.

I am sure that this ISEMMS will be the start of a successful series of conference events where the PC activity can be worked out in the visibility of its achievements in coordinating the IFToMM community in the several aspects of MMS education, as asked during my IFToMM Presidency. Nowadays more than in the past, there is a need to share and discuss the current problems in MMS education with the aim to give solutions and to identify proper trends for a modern worldwide common vision that will consider MMS still a core area in Engineering formation within Technology developments for Society enhancement.

This is to congratulate the organizers and authors for this book giving an overview of problems and situations on MMS Education with several points for future discussions and possibilities of improvements of our future.

Leganés, June 2013 Marco Ceccarelli

Preface

Contents

Introduction

The First International Symposium on the Education in Mechanism and Machine Science (ISEMMS 2013) is the first event of a promising series of symposiums focused on Mechanism and Machine Science (MMS) education. In the 1970s of the last century, Prof. Artobolevsky proposed a first seminar about 'Education in MMS'. Today, many years after this pioneer seminar, the Permanent Commission (PC) of Education of IFToMM has established these new series symposium. This first conference has been held at Madrid on June 13–14 of 2013.

The aim of the ISEMMS symposium is to establish a forum where teachers and researchers interchange experience in the field of MMS.

We have received 44 abstracts by authors from all around the world. After the review process 32 contributions were selected for presentation and publication in this book. The program of ISEMMS 2013 has included technical sessions with oral presentations, two plenary sessions and social activities that enhanced to share experiences, discussions and let all the participants to start new collaborations among them. The contributions cover topics related with new trends and experiences, mechanical engineer curricula, methodology and virtual labs.

We would like to express grateful thanks to the members of the steering committee for ISEMMS 2013 who have collaborated for the success of the symposium, especially Professors Pietro Fanghella, Joerg Bauer and Olga Egorova.

We thank Universidad Carlos III of Madrid (UC3M), IFToMM, AEIM (Spanish Association of Mechanical Engineering) and the Mechanical Department of UC3M for the financial support for the ISEMMS 2013.

We thank to the members of MAQLAB research team for the great effort they have made during the preparation and holding of the symposium. They have done a very good job.

We also thank Natalie Jacobs and Anneke Pot (Editorial board of Springer) for the editorial support of this book.

And finally, we would like to acknowledge the invaluable support of Marco Ceccarelli during the first stages of this symposium and the continuing help to achieve with success the ISEMMS 2013.

We thank the authors for their excellent and interesting contributions. We think the works presented in this book will be of interest for the MMS educators community and the ideas, experiences and developments presented shall enhance the quality in this main area.

Juan Carlos García-Prada
Cristina Castejón

Part I
Mechanism and Machine Science in the Mechanical Engineer Curricula

Historical Accounts on the Figure of Engineers and Academic Mission for their Formation

Marco Ceccarelli and Roberto Bragastini

Abstract The mission of academic formation of mechanical engineers is strongly connected with role of their profession in the society. This has evolved from practitioner on field to a science teaching through a slow evolution. This paper is an attempt to outline the main aspects of historical developments as well as current critical issues in recognizing a significant role to mechanical engineering and its academic formation. In particular, the paper presents an historical evolution of the understanding of engineers figure and consequently it outlines the mission issues in a modern academic formation.

1 Introduction

Today the figure of engineers suffers of considerable reduction of significance and influence in the society when compared with respect to the great reputation in the past. Even more is the reduction of authority for mechanical engineers.

What are the motivations of such a less attraction to science and technology in this modern time that nevertheless is dominated by the success of engineering?

The problem has been attached and still is attached at several levels and with different views in these last years and a considerable literature is already published on the topic. Strong attention is address within formation (University) frames and professional applications, even with discussion of reform proposals.

M. Ceccarelli (✉)
Laboratory of Robotics and Mechatronics, University of Cassino and South Latium, Cassino, Italy
e-mail: ceccarelli@unicas.it

R. Bragastini
Independent Scholar (Retired Professor from Padua University), Via Piave 3/20 , 95027 San Gregorio di Catania, Italy

J. C. García-Prada and C. Castejón (eds.), *New Trends in Educational Activity in the Field of Mechanism and Machine Theory*, Mechanisms and Machine Science 19,
DOI: 10.1007/978-3-319-01836-2_1,

This paper is an attempt to explain the current situation by looking at the historical evolution of the figure and task of engineers as understood within a technical community whose authors are members with different experiences and perspectives. Thus, the authors have discussed and illustrated the significance and success of engineers though significant examples over the time, with the aim to stress the cultural importance of engineers in the society with a vision for better life quality as consequence of a proper merit and consideration of engineers activities.

2 On the Term of Mechanical Engineer

The figure of mechanical engineers can be understood when mechanical issues are understood by looking at the corresponding terminology that can be summarized by referring to IFToMM definitions [6, 11, 12]:

- Machine: Mechanical system that performs a specific task, such as the forming of material, and the transference and transformation of motion and force.
- Mechanism: System of bodies designed to convert motions of, and forces on, one or several bodies into constrained motions of, and forces on, other bodies.

A modern term MMS (Mechanism and Machine Science) has been adopted within the IFToMM Community to better indicate the complexity of modern machines as:

- Mechanism and Machine Science: Branch of science, which deals with the theory and practice of the geometry, motion, dynamics, and control of machines, mechanisms, and elements and systems thereof, together with their application in industry and other contexts, e.g. in Biomechanics and the environment. Related processes, such as the conversion and transfer of energy and information, also pertain to this field.

Interesting is the evolution of machine definition that can be summarized thought the following significant examples:

- By Marcus Pollione Vitruvius (he lived in first century B.C.) in De Architectura-liber X, translated and discussed by Daniele Barbaro in [2]: “A machine is a combination of materials and components that have the capability of moving weights”;
- By Galileo Galilei in Le Mecaniche [8]: “A machine is a means by which a given weight will be transported to a given location by using a given force”;
- By Paolo Branca in [4]: He described machines by stressing “the operation as consumption of motor power and changes of weights in functions of time history, cost and operator skill”;
- By Jacob Leupold in [16]: He treated the description of machines and mechanisms referring to “their aim of modifying motion rather than just the construction of machinery”;

- By Josè Maria de Lanz and Augustin de Betancourt in [15]: "In agreement with Gaspar. Monge, we consider as elements of machines the devices than can change the direction of the movements…the most complicated machines are only combinations of those capable of single movements";
- By Robert Willis in [22]: "I have employed the term Mechanisms as applying to combinations of machinery solely when considered as governing the relations of motion. Machinery as modifier of force";
- By Franz Reuleaux in [19]: "A machine is a combination of bodies capable of withstanding deformation, so arranged as to constrain the (mechanical) sensible forces of nature to produce prescribed effects in response to prescribed input motions";
- By Francesco Masi in [17]: "Hence we name: as mechanism a kinematic chain that has been fixed on one of its components; as machine a mechanism whose components make mechanical work";
- By Gabriel Koenigs in [14]: "A machine is recognized as an assembly of resistant bodies that are constrained reciprocally and are under the action of natural forces. If you abstract from forces, the remaining of a machine consists of bodies and constraints. This is a mechanism";
- By Richard S. Hartenberg and Jacques Denavit in [10]: "The term machine is associated with the use and transformation of force, and although motion is varying degree is encountered in a machine, the idea of force dominates. Mechanism, on the other hand, definitely conjures up the idea of motion, and while forces do exist, they are relatively small and unimportant compared with the exploitation of motion";
- By Joseph Edward Shigley and John Joseph Jr Uicker [13]: "A machine is an arrangement of parts for doing work, a device for applying power or changing its directions. It differs from a mechanism in its purposes. In a machine, terms such as force, torque, work, and power describe the predominant concepts. In a machine, though it may transmit power or force, the predominant idea in the mind of the designer is one of achieving a desired motion";
- By IFToMM, the International Federation for the Promotion of Machine and Mechanism Science [12]: "A machine is a mechanical system that performs a specific task, such as the forming of material, and the transference and transformation of motion and force; and a mechanism is a system of bodies designed to convert motions of, and forces on, one or several bodies into constrained motions of, and forces on, other bodies".

More specifically the meaning and role of (mechanical) engineers have evolved from the different perspectives.

Today in the Oxford dictionary [18] an engineer is defined from historical viewpoint as 'a designer and constructor of military works' and from general understanding as 'a person whose occupation is the design, construction and maintenance of works of public utility'. Similarly, from Italian UTET dictionary [21] in antiquity an engineer was a person, who designed and built any kind of devices and machinery, with particular mention to war machines and hydraulic systems. Today from general understanding an engineer is a person, who with a formation on science and technology works for design, construction, and operation of systems, [9].

From the above few examples (and more can be noted as from other countries) it is remarkable that still today there is not a well focused understanding of the figure of engineers with a well identified role and activity, although is universally recognized the specific activity of engineers in developing systems for the needs of the society.

From legal viewpoint the situation is even more confused and still under definition, both at national and international levels, since the today large variety of engineering fields and their areas of implementations both in scales and natures of the objects they refer to. Other situations have complicated the situations. For example historically in Italy the legal definition of the figure of engineer has suffered a fragmentation due to the several kingdoms that were persistent for centuries and only the unification in 1861 imposed a need of a unique legal position that nevertheless was not easy to achieve and it still under reforming actions, due to also European constraints beside professional prerogatives and perspectives even on local conditions.

An understanding of the role and activity of engineers can be understood by looking at the epistemology of the term engineer. A short account is given in the following.

The term engineer was coined through an evolution from a word in Latin that was used at the end of the Roman empire at the end of fifth century up to a modern meaning that was established at the time of the Ecòle Polytechnique in Paris at the beginning of nineteenth century. Technical formation of engineers, as independent from military corps, was a need felt around the world for the demand of engineers in the developing technology since the early stages of Industrial Revolutions.

A specific term as 'ingeniator' appeared in eighth and ninth century in south France and north west Italy to indicate the profession of builders of apparatus. Just later the term evolved to 'engineor' and 'engineur' in south France, where as in north Italy it became 'encignarius'. In particular, a first mention of the term can be found in a legal act in Genova, Italy with the date of 19 April 1195 where the role of a cited Rainaldus is named as 'encignerius', [20]. Relevant is also an edict by Ludovico il Moro, duke of Milan who in 1492 prescribed that civil constructions must worked out by 'ingeniarii'.

3 Role and Activity of Mechanical Engineers

A historical survey of the role and activity of (mechanical) engineers over the time can be outlined by looking at emblematic personalities, but also without forgetting the many anonymous persons. In the following a short survey is reported to indicated main characters over the time.

Archimedes (287–212 B.C.) can be considered a first machine engineer with a very strong scientific activity [7], by combining theory (investigations) and practice (applications). Because of both aspects he had a great reputation also in the centuries after his death.

Francesco di Giorgio (1439–1501), although architect, was the typical Renaissance engineer. He dedicated lot of successful efforts in designing, constructing, and operating a large variety of machines for the enhancement of production and help of human labour, including automata for entertainment. Most of his machine designs were improvements of previous solutions, but several ones were products of his ingenuity with long-duration engineering source.

James Watt (1736–1814) is considered the first modern mechanical designer engineer since he improved and indeed made powerful the steam engine with proper mechanism designs completing an efficient motion and power transmission, with an empirical ingenuity that was later the base for advances in theory and practices of mechanisms and machines.

Augustin Betancourt [12] can be considered the first modern manager mechanical engineer since he used his knowledge and expertise on mechanical systems both in theory and design in developments for Industry frames in several countries (Spain, Frances, Russia) to give a supranational impression of the figure and role of engineers, with a great reputation and influence.

Franz Reuleaux [19] can be considered an emblematic example of modern scientist in MMS, who studied the many possibilities of mechanisms for machine design both by developing a general theory and applying it in practical industrial applications.

Ivanov I. Artobolevski [1] can be considered a modern mechanical MMS engineer, who has centred his activity on Mechanism Design to elaborate the successful most wide mechanism handbook as results of his activities in practical engineering based on theoretical works. His success as mechanism designer and inventor gave him a great reputation and role in many frames of the society both at national and international levels.

Corradino D'Ascanio (1891–1981) can be considered an emblematic example of engineer inventor of machines of current technology, since his creativity and MMS formation gave him the possibility to design and produce brilliant market-successful machines, like the scoter Vespa and first helicopter.

The above are examples of figures with significant activity in engineering with an influential role in the developments in Technology and finally in the society. But they have reached those goals also thanks to a multitude of anonymous engineers whose works has prepared and indeed stimulated those achievements. A specific large literature to which a reader can refer, is available, even with several different perspectives, on each of the above personalities and many more, with details of their legacy and contributions on technological developments and education enhancements.

4 On the Formation of Engineers and Mission of University

Today professional mechanical engineers are formed, even with a time sequence, within the frames of University, professional Unions, and enterprise training. Engineers receive formation with mainly information and theoretical backgrounds during University curricula; specific profession-oriented formation is offered within activities of engineers Unions; and finally employment-oriented training is given by industry while searching for young engineers. Recently, plans are worked out also for so-called continuous education that permits engineers to be updated on the advances of Technology and their methodologies by means of well-focused short courses or teaching programs within University and/or Union frames.

But academic teaching can be still considered a fundamental base for a scientific/technological formation of young successful engineers, where as the above-mentioned formation activities at engineers Unions and company training centres can be considered complementary or very specifically focused on a strict area of employment.

In general, the current formation of engineers is planned with three main parts of teaching, namely fundamentals of science, basic engineering, and technology for engineering, as by a tradition coming from the technical schools of the nineteenth century. The fundamentals of Science are given to provide the means and knowledge of using Science. The basic Engineering gives the methodologies and approaches for designing and developing ideas and systems, where as Technology teaching for engineering is aimed at providing the means and applications of implementing engineering designs. These frames of academic teaching are finalized to provide skills and knowledge to attach problems for current but innovative solutions. The mission of University frames is ultimately aimed at disseminating the accumulated/acquired knowledge by proving new generations of engineers with a well aware capability of implementing those knowledge in practise both in Industry and Society. Both handling of traditional technology and developing innovation are fundamental for an application of technology towards benefits to Society for better life of mankind.

Today there is great attention towards innovation, perhaps because the accelerated evolution of Science requires a similar output in Technology. In general innovation is motivated only by a need of market prominence with products that being the first offer or with a cheaper cost can ensure success to the company investment more than its manufacturing or even society enhancement. But what is really Technology innovation? Do we really need to innovate and forget about the well established developments? Why innovation cannot be considered even as further reconsideration and improvement of existing current successful solutions? Finally, form conceptual viewpoint is it really new what is presented as innovation?

In general, engineers of the past have achieved innovation though inventions but improvements of machinery with a (relatively slow) evolution of performance in design and operation of new and old tasks/needs [3, 5]. Most of the past achievements were forged within or from academic frames with more and more collaborations as intended to fulfil the above-mentioned universal mission of Universities. Is this still possible today?

Universities need more international links not only in research activities for producing that accumulation of knowledge, but mainly for providing frames of so-called global formation that makes young engineers ready to work out both current Technology and innovation at worldwide level, but in any frames even with different cultural surroundings. International bodies, like IFToMM (The international Federation for the Promotion of Mechanism and Machine Science) as the only worldwide society in mechanical engineering, can help in such a mission for the specific focus on specific areas by aggregating a community with common views and targets. But what is expected from this activity by a general public understanding to recover a reputation for engineers and their formation?

The past reputation of engineers and technology scientists with significant consideration in the society (since engineers often became leaders in national governments) was based and indeed recognized in the fact that their work and achievements gave the possibility of enhancing the society and ultimately the quality of life of all citizens. Why today we lost such a reputation and consideration, although the society improvements still depend form technological developments, as clearly experienced in the past toe decades?

But what happened and what is happening why Academic frames are loosing the reputation and consequently the support of Society in the mission of formation of new generations of engineers. Is this only matter of costs?

Perhaps the mission of University needs to be reshaped as asked by political governors with different approaches and better efficiency.

But nevertheless the above perspectives and considerations of academic formation can be considered useful, when giving the knowledge of Science and Technology towards skills in handling the current but innovative solutions.

Finally, we believe that University still has and must have the mission of proper wide formation of next generations of engineers by using the accumulation and development of Science and Technology. In fact, a role of cultural significance even with long-term goals can be still vindicate to Universities, like in the past, as being the only frames in which the future can be shaped with new visions in the current but future generations.

5 Conclusions

The figure of (mechanical) engineers has evolved from practitioners with empirical experiences to professionals of Science applications with significant roles in the society with the aim of enhancing the conditions of better quality of life and labour.

Correspondingly the formation has been worked out with direct transmission of knowledge and skills within limited group of persons up to wide dissemination by the modern University frames with further specialization through professionals Unions and enterprise employments.

The mission of University formation that is reinforced by IFToMM in the specific area of MMS, can be still understood as aimed at producing advances through research activities and promoting developments but implementations for the benefit of society. But unfortunately such a fundamental task does not correspond today with a social reputation of engineers in the current society like in the past. What and how should the mechanical engineers recover or enhance their formation and activity in order to re-gain a significant reputation, role, and finally leadership in society developments?

References

1. Artobolevsky, I.I.: Mechanisms in Modern Engineering, vol. 5. Mir Publishers, Moscow (1975–1980)
2. Barbaro, D.: I Dieci Libri dell'Architettura di M. Vitruvio. Appresso Francesco de' Franceschi Senese, Venezia (1584)
3. Besson, J.: Théatre des Instruments Mathématiques et Mécaniques. Barthélémy Vincent, Lion (1578)
4. Branca, G.: Le Machine. Ad ista[n]za di Iacomo Martuci ... per Iacomo Mascardi, Roma (1629)
5. Bresse, J.A.: Cours de Mecanique et Machines. Ecole Polytecnique, Paris (1885)
6. Ceccarelli, M.: On the meaning of TMM over time. Bull. IFToMM Newsl. **8**(1) (1999)
7. Ceccarelli, M.: The mechanics of Archimedes towards modern mechanism design. In: Paipetis, S.A., Ceccarelli, M. (eds.) The Genius of Archimedes: 23 Centuries of Influence on Mathematics, Science and Engineering, Book series on History of Machines and Machine Science, vol. 11, pp. 177–187. Springer, Dordrecht (2010)
8. Galilei, G.: Le Meccaniche. reprinted in Opere di Galileo Galilei Edited by Brunetti, F., Torino, 1964 (1593)
9. Garzanti: Dictionary of Italian language. Garzanti, Bologna (2010)
10. Hartenberg, R.S., Denavit, J.: Kinematic Synthesis of Linkages. Mc Graw-Hill, New York (1964)
11. IFToMM: IFToMM commission a standard for terminology. Mech. Mach. Theory **26**(5), 435–539 (1991)
12. IFToMM: IFToMM commission a standard for terminology. Mech. Mach. Theory **38**(7–10), 598–1111 (2003)
13. Shigley, J.E., Uicker, J.J.: Theory of Machines and Mechanisms, 2nd edn. Mc Graw, New York (1995)
14. Koenigs, G.: Introduction a une Théorie Nouvelle des Mécanismes. Librarie Herman, Paris (1905)
15. Lanz, J.M., Betancourt, A.: Essai sur la composition des machines. de L'Imprimerie Impériale, Paris (1808)
16. Leupold, J.:Theatrum Machinarum. Christoph Zunkel, Leipzig (1724)
17. Masi, F.: La teoria dei meccanismi. Zanichelli, Bologna (1897)
18. Oxford Dictionary 1988, Dictionary, London (1988)
19. Reuleaux, F.: Theoretische Kinematic. Vieweg und Sohn, Braunschweig (1875)
20. Treccani 2000: Encyclopedia, Torino (2000)
21. UTET: History of Technology. UTET, Turin (1974)
22. Willis, R.: Principle of Mechanisms, 1870–2nd edn. Wiley, London (1841)

Toward a Curriculum in Mechatronics: Two Experiences in Italy and Spain

F. Bonsignorio, L. Bruzzone and P. Fanghella

Abstract The paper carries out the comparative analysis of two bachelor programs focused on mechatronic and system engineering in two Universities in Italy and Spain. Two levels of analysis are considered: at program level, the respective structures are analyzed and compared, also taking into account the academic origin of each program; then a similar specific course for each program is discussed and different teaching and student evaluation methods compared. Final considerations and conclusions complete the analysis.

Keywords Mechatronics · Mechanical engineering · System engineering and automation · Bachelor

1 Introduction

Mechatronic engineering is certainly a relevant field in contemporary industrial and information engineering. Starting from early 1970s [1], it has grown in importance, and today mechanical systems, with embedded sensor, control and actuations (i.e., mechatronic systems) play a key role in our everyday life as well as in complex industrial environments. The intrinsic interdisciplinary nature of the mechatronic field has caused a certain difficulty in the development of suited curricula in academic organizations. As a matter of fact, the implementation of interdisciplinary tracks mainly require an "horizontal" point of view crossing the boundaries of academic organizations that are historically organized with a strong "vertical" disciplinary approach.

F. Bonsignorio (✉)
University Carlos III of Madrid, Madrid, Spain
e-mail: fabio.bonsignorio@uc3m.es

L. Bruzzone · P. Fanghella
DIME—University of Genoa, Genoa, Italy
e-mail: bruzzone@dimec.unige.it

J. C. García-Prada and C. Castejón (eds.), *New Trends in Educational Activity in the Field of Mechanism and Machine Theory*, Mechanisms and Machine Science 19,
DOI: 10.1007/978-3-319-01836-2_2,

The problem has been studied by several authors, with different perspectives: from an organizational academic point of view [2–4], discussing the state of mechatronic curricula in specific Universities [5–10], and with a wider perspective, analysing the situation of mechatronic education with reference to national or large geographic areas [11–16]. As shown in references, the number of programs in mechatronic engineering is rapidly growing, often due to initiatives of cultural groups that, involving other interested areas, set up the curricula.

The goal of the paper is to analyse and compare the characteristics of two different first level degrees that, albeit with different names and origins can be both located within the mechatronics and automation area. The two programs are held in two primary Universities—University Carlos III in Madrid, Spain, and University of Genoa, Italy—within their respective Schools of Engineering. It is interesting to observe that the two countries have interpreted in slightly different ways the Bologna Declaration [17] stating the three level organization of university paths: as discussed in next Sections, for the first two levels, the Italian curricula are based on the so called $3 + 2$ year scheme, with a bachelor degree corresponding to 180 ECTS credits (called "Laurea") followed by a master degree lasting 2 years (120 credits) [18]. The Spanish scheme is more flexible, and is based on a first level degree (called "Grado") lasting 4 years (240 credits) followed by a Master track corresponding to an effort variable between 1 and 2 academic years (60–120 ECTS credits).

2 Outline of Curriculum in Italy

The course considered for the analysis carried out in the paper is a first level degree named "Mechanical engineering—Automation and Mechatronics curriculum". The curriculum comes out as a spin-off of a general first level degree in Mechanical Engineering held in Genoa, and is taught in the separate campus in La Spezia. Table 1 shows the structure of the curriculum; for all courses it is shown the number of ECTS credits and the cultural area of the course (Base = general math, physics or other introductory courses in different scientific fields; EEI: Electrical, electronics and computer engineering, ME: Mechanical engineering field, SEA: System engineering and automation).

A careful analysis of the curriculum structure reveals that it comes out from the ME context: in particular this is evidenced by the presence of a relevant amount of credits in traditional ME areas such as: Materials, Mechanics of machines, Manufacturing, Heat transfer, Energy systems and Turbo machinery, Design of machines. Moreover also the boundary areas treated in the "Dynamics and Control of mechanical systems" and "Measurements and Instrumentation" courses are here attributed to the ME field, and are taught from lecturers belonging to such cultural area. Beyond the traditional mechanical contents, the mechatronics track is implemented through two paths of 18 ECTS credits each, one mainly oriented to electronics and electrical drives, the second to software development and SEA. Time limitation due to the three year organization has strongly constrained the design of the curriculum, limiting the complexity of

Table 1 Courses of mechatronic engineering curriculum—University of Genoa

Subject	Credits	Area
Calculus 1	12	Base
Chemistry and engineering materials	12	Base
English language	3	Base
Geometry and linear algebra	6	Base
Physics 1	12	Base
Computer and programming	6	EEI
Technical drawing	6	ME
Calculus 2	6	Base
Mechanics of machines	12	ME
Manufacturing and industrial plants	12	ME
Electric and electronic systems	12	EEI
Thermodynamics and heat transfer	12	ME
Energy systems	6	ME
Computer and programming 2	6	EEI
Design of machines	12	ME
Dynamics and control of mechanical systems	6	ME
Electric drives	6	EEI
Embedded systems	6	SEA
Fluid machinery	6	ME
Measurements and instrumentation	6	ME
Apprenticeship and final exam	3	
Electives	12	

the presented knowledge and the skills to be acquired. This problem is particularly relevant for a curriculum that presents an interdisciplinary nature, as the large set of topics and disciplines to be considered either requires a large and long learning effort or as said before, it poses strong limitations to the taught matter.

As a final consideration, it is worth noting that the introduction of the mechatronics curriculum as an enrichment of the ME course at La Spezia campus of the University of Genoa has corresponded to a doubling of the number of students enrolled to the course, while, for the corresponding academic year, the University has suffered a global reduction of about 10 % of the first year students.

3 Outline of Curriculum in Spain

The Spanish curriculum considered for the comparison is the Bachelor in Automation and Industrial Electronic Engineering. The bachelor and masters in Industrial engineering, in Robotics and Automation, in Aerospace Engineering aim to cope with the needs of global class manufacturing from companies such as EADS and John Deere, which have important manufacturing sites in Madrid area, and other firms of the emerging high tech industry.

This curriculum is focused on Mechatronics, i.e., on system engineering, automation and robotics issues. The professional opportunities for a graduate in Automation and Industrial Electronics Engineering are varied, they span from the design and development of system integration and automation projects, to domotics, to design of electronic devices, instrumentation, laser devices, robotics, etc.

The degree has a deliberately significant practical component. The employment rate for graduates in Automation and Industrial Electronics Engineering has been consistently comparatively high, with a low average time to find a job. Usually more than 95 % of the graduated are employed one year after graduation, although this may have worsened in the last year due to the well-known external conditions. Table 2 shows the structure of the curriculum; for all courses it is shown the number of ECTS credits and the cultural area of the course (SEA: System engineering and automation; Hum: Humanities).

Elective Group 1 includes: Industrial Automation II, Electromechanical Actuators, Intelligent Control, Real time systems, Systems of Perception, Computing Systems, Simulation of dynamic systems, Microprocessors, Robotics , Optelectronics, Electronic Instrumentation II, Analog Electronics II, Integrated circuit design, Manufacturing and construction of electronic equipment, Electronic system design, Professional Internships. Elective Group 2 includes: Home and building automation, Vehicle Automation, Control and automation applications in biomedicine, Control Engineering III, Digital electronic systems, Power electronics systems, Electrooptic Systems, Microelectronics, Professional Internships.

The students must choose 4 courses (24 credits) from Elective group 1 and 2 courses (12 credits) in Elective group 2. Among the elective subjects, a student can choose an internship in a company for 6 credits. Students who pass the entrance exam in English have the possibility to attend more than 80 % of the courses in English. In this curriculum (240 ECTS in 4 years), the control, industrial automation and robotics, and Electronics technology, themes account for a significant part of the ECTS credits as shown in Table 2. The availability of 4 years allows, at least in principle, an higher level of knowledge and skill at graduation and it is somehow in the middle between the traditional Italian and Spanish degrees and the 3 years grade of the 3 + 2 system implemented in Italy. The presence of mandatory courses in Humanities is a peculiar characteristic of Carlos III.

4 Detailed Analysis of a Course in Italian Curriculum

The course chosen for a detailed comparative analysis is an introductory course regarding the dynamics of linear systems and their control, named “Dynamics and control of mechanical systems”. The course is located in the first semester of the 3rd year, and corresponds to six ECTS credits. It is preceded, in the 1st and 2nd years, by the classical math courses (linear algebra, differential equations) and by a course in Mechanics of machines, that introduces the students to the dynamics of single d.o.f., constant transmission ratio mechanical systems and to single d.o.f.,

Table 2 Courses of mechatronic engineering curriculum—Carlos III University— Madrid

Subject	Credits	Area
Calculus 1	6	Base
Linear algebra	6	Base
Physics I	6	Base
Programming	6	EEI
Information skills	3	Hum
Writing and communication skills	3	Hum
Calculus II	6	Base
Chemical basis of engineering	6	Base
Engineering graphics	6	ME
Physics II	6	Base
Statistics	6	Base
Materials science and engineering	6	Base
Industrial automation I	6	SEA
Electrical power engineering fundamentals	6	EEI
Machine mechanics	6	ME
Thermal engineering	6	ME
Fluid mechanics	6	ME
Introduction to engineering management	6	ME
Mechanics of structures	6	ME
Electronics engineering fundamentals	6	EEI
Production and manufacturing systems	3	SEA
Environmental technology	3	Base
Electrical machines and installations	6	EEI
Control engineering	6	SEA
Computing systems	6	SEA
Digital electronics	6	EEI
Analog electronics I	6	EEI
Control engineering II	6	SEA
Industrial robotics	6	SEA
Electronic instrumentation I	6	EEI
Power electronics	6	EEI
English language	6	Hum
Humanities	6	Hum
Electives (Group 1)	24	
Technical office	3	EEI
Industrial organization	3	ME
Bachelor thesis	12	
Electives (Group 2)	12	

vibrations. Moreover, they will have a 6 + 6 credits course in electrical and electronic engineering fundamentals. Prior to this course, they do not have any systematic background in linear system dynamics.

The course is aimed at teaching the students a few basic, but founding, concepts that are deemed as fundamental in modern engineering. Moreover, to deepen student knowledge, the acquisition of operative skills implemented through the Matlab software and its Control System Toolbox is required. Due to its introductory nature, the course is based on the study of the properties of SISO systems, through the Laplace transform notation and the transfer function approach. No state space representations are introduced and discussed.

According to its name, the first part of the course presents an introduction to linear system theory, regarding the following main topics: introduction to SISO-LTI dynamic systems, with examples focused on mechanical and electro-mechanical systems, properties and use of Laplace transform for the analysis of LTI systems, transfer functions, time response (impulse, step) frequency response, Bode plots and their relation with system properties. In particular, the comprehension of the mathematical and physical meanings of frequency response is considered a key goal of the course, as this way of studying system properties plays a very relevant role also in other areas such as instrumentation and measurements and mechanical vibrations.

The second part, more or less occupying two credits, is devoted to the study of closed loop systems and to the solutions needed to obtain nice performances from them, the main considered topics being: closed loop transfer function, criteria to define the performances of servo-systems (stability, precision and speed) and their relations with open-loop frequency response, main regulator types, choice of regulator and of its numerical parameters for prescribed closed loop performances.

All course lectures are given in computer lab, and the lecturer can freely mix the presentation of theory with practical exercises carried out analytically or with the aid of Matlab. All along the course very strong emphasis is put on the relation of the presented math results, that occupies a large part of the course, with their physical meanings and interpretation. The course does not introduce intermediate tests, and, very differently from its Spanish counterpart (see next Section) does not have written examinations. The final exam is organized as a discussion in which the student is required to discuss the properties of a given dynamic system (e.g., stability, step response, frequency response) representative of a "plant" in the second part of the exam the control of the plant is analysed, usually starting with a unitary regulator, and then adding proper control actions according to given specifications. The exam can be failed at any moment in case the student is not able to show the comprehension of some basic ideas, such as, for example, the concept of frequency response or the meaning of phase and gain margins.

5 Detailed Analysis of a Course in Spanish Curriculum

The course considered for the comparison is an introductory course in control engineering, called Control Engineering I. This course is situated in the first semester of the 3rd year and is preceded by basic math courses similar to those imparted in the course at the University of Genoa. Actually some of the mathematical techniques exploited in the course are only partially developed in previous teachings, in particular the methods of complex field calculus. This teaching aims to give a substantial understanding of basic control engineering and theory concepts and methods. It is focused on SISO (Single Input Single Output) and LTI (Linear Time Invariant) systems. The design methods are restricted to PID systems.

In particular the students are introduced to the concept of 'dynamical (linear) system', Fourier and Laplace transforms, time and frequency response of LTI systems, stability analysis methods, PID design methods, in particular the Root Locus method, Nyquist and Nichols methods. In practice the course is articulated on three pillars, a set of 'Theory' lectures, 'Problem Solving' and Lab sessions, consisting of various problem solution examples with Matlab. Matlab examples are actually also shown during the 'theory' and problem solving lectures, but in this case the students do not have individual personal computers. The lectures are, usually, but not always, depending on the number of students and teachers' workload, given by three distinct pools of professors. The evaluation of the students is continuous, although the students need to attend a final exam (accounting for 40 % of the final mark). The exams are usually written for quality assurance reasons. In particular it is customary to share the preparation and sometimes to cross the evaluation of the final exam among the 'theory' professors. The students also have to solve a number (6–8) of homework under time pressure and to attend two continuous evaluation exams during the semester. The exams (and the exam questions and exercises) are usually evaluated individually and the evaluation criteria are known to the students from the beginning of the year. As a consequence a student can fail (even with a zero) in a question or exercise of a test and pass it.

6 Discussion

The comparative analysis of the two curricula reveals several interesting differences, most likely caused by the cultural groups originating the two programs, but impacting the education paths of the students. More specifically, as outlined in Table 3, although one year shorter, the Genoa curriculum has a wider contributions content in the field of ME, also due to several boundary topics dealt by the ME professors. On the other side, SEA and EEI areas are strongly represented in Spanish program; also resources devoted to electives and humanities (foreign language and humanities) are significantly higher in the Spanish program.

On the one hand the design of curriculum of studies with similar objectives have been carried out with some significant differences in the two universities we consider, on the other hand the two introductory courses on control engineering are very similar from the content stand point. Both present an introduction to linear system theory, focusing on SISO-LTI dynamic systems, both analyse examples of mechanical and electro-mechanical systems, introduce Laplace transform for the analysis of LTI systems, transfer functions, time response (impulse, step, ramp), frequency response, Bode plots and their usage for system analysis. There are some differences. In Genoa,

Table 3 Credit distribution for the two programs

Program		Base	Eei	Sea	Me	Electives
Genoa	credits	51	30	6	78	15
	% total	28 %	17 %	3%	43 %	8 %
Madrid	credits	69	51	33	39	48
	% total	29 %	21 %	14 %	16 %	20 %

frequency response is given more importance as this approach is quite useful in the study of instrumentation and measurements and mechanical vibrations. Both courses introduce design methods of PID systems, and the basic concepts and methods of phase margin and stability margin. In Genoa the analysis is mainly carried out by Bode diagram methods, in Madrid, the root locus method and Nyquist's diagram are also introduced. In both Universities the presentation of the topics is relatively coarse from the mathematical standpoint and, in the engineering traditions, aims to stimulate the intuition of the students, rather than giving absolutely rigorous mathematical demonstrations. Where the two courses differ more in the evaluation procedure: in Genoa only oral exams are conducted, while in Madrid there is a more complex continuous evaluation process based on a mixture of intermediate tests, homework under time pressure and lab sessions. In Madrid oral exams are an exception. Another difference is that in the oral exam in Genoa there are 'showstopper' questions, which make the student fail whatever the student's answer to the other questions. In Madrid the evaluation is continuous along the whole teaching, while in Genoa it is at the end. An anecdotal difference is that in Genoa the marks are in the range 0–30, the student pass with 18, in Madrid from 0 to 10 with the threshold set at five. It would be interesting to find ways to objectively verify which evaluation process is more accurate and productive for the students.

References

1. More, T.: Mechatronics, Yaskawa Internal Trademark Application Memo 21.131.01, July 12 (1969)
2. Grimheden, M., Hanson, M.: Mechatronics: the evolution of an academic discipline in engineering education. Mechatronics **15**, 179–192 (2005)
3. Das, S., Yost, A.S., Krishnan, M.: A 10-year mechatronics curriculum development initiative: relevance, content, and results: Part I. IEEE Trans. Educ. **53**(2), 194–201 (2010)
4. Krishnan M., Das S., and A. Yost A. S.: A 10-year mechatronics curriculum development initiative: relevance, content, and results: Part II. IEEE Trans. Educ. **53**(2), pp. 202–208 (2010)
5. Wang, Y., Yu, Y., Xie, C., Wang, H., Feng, X.: Mechatronics Education at CDHAW of Tongji University: Laboratory Guidelines. Framework Implementations Improvements Mechatronics **19**, 1346–1352 (2009)
6. Wang, Y., Schwarz, B., Pan, Y., Chen, B., Wiedmann, H., Xie, C., Wang, H., Yu, Y., Feng, X.: Mechatronics education at CDHAW of Tongji University: Structure. Orientation Curriculum Mechatronics **18**, 172–177 (2008)
7. Tutunji, T.A., Jumah, M., Hosamel-Deen, Y., Rabbo, S.A.: Mechatronics curriculum development at Philadelphia University in Jordan. Mechatronics **17**, 65–71 (2007)

8. Brown, N.J., Brown, O.T.: Mechatronics a graduate perspective. Mechatronics **12**, 159–167 (2002)
9. Wright, A.B.: Planting the seeds for a mechatronic curriculum at UARL. Mechatronics **12**, 271–280 (2002)
10. Allen R. G.: Mechatronics engineering: a critical need for this interdisciplinary approach to engineering. In: Proceedings of The 2006 IJME—INTERTECH Conference (2006)
11. Lee, J.W.: Investigation of mechatronic education in South Korea. Mechatronics **20**, 341–345 (2010)
12. Akpinar, B.: Mechatronics education in Turkey. Mechatronics **16**, 185–192 (2006)
13. Loose, H.: Mechatronics engineering programs at German Universities of Applied Sciences. Solid State Phenomena **65**, 419–424 (2010)
14. Tan, K.K., Lee, T.H., Dou, H.F., Lim, S.Y.: Various developments in mechatronics in Asia. Mechatronics **8**, 777–791 (1998)
15. Bradley, D.: What is mechatronics and why teach it? Int. J. Electr. Eng. Educ. **41**, 275–291 (2004)
16. Fanghella P.: Education in mechatronic engineering – Italian perspective. In: 13th World Congress in Mechanism and Machine Science, Guanajuato, México, 19–25 June (2011)
17. The Bologna declaration, Joint declaration of the European Ministers of Education. http://www.ehea.info/ Accessed 19 June 1998
18. ECTS Users' Guide, European, Communities, 2009 (ISBN 978-92-79-09728-7)

Mechanical Engineering at a Distance: A Review

M. Artés and J. López

Abstract Distance learning has experienced a great development in the last decades due to the advances achieved in the field of information and communication technologies and also because of sociopolitical reasons, being encouraged by governmental policies and offering a chance to people that because of economic, working, geographic or residential reasons would not have any other way to access to university studies. Under these circumstances there has been a huge proliferation of distance education programs all throughout the world and mechanical engineering has not been an exception. The academic offer is not restricted to universities specialized on distance teaching methodologies, such as open universities or distance learning universities, since most universities based on traditional educational methodologies now incorporate distance education on their academic programs. However, it is not frequent to find a mechanical engineering degree where the whole program is based on distance learning. This is primarily due to the need that students develop laboratory practices. Although some mechanical engineering lab practices can be made at home with special kits and virtual or remote labs may be viewed as an alternative, most universities request compulsory stays at their facilities. This is the rationale why it is not very common to include mechanical engineering degrees in the technological academic offer of a lot of universities based only on distance learning. In this work the current status of mechanical engineering education at university level in the frame of distance learning is presented, analyzing its concept and evolution, global presence, methodologies and educational resources.

Keywords Mechanical engineering · Distance learning · Academic programs · Educational resources

M. Artés (✉) · J. López
University UNED, Madrid, Spain
e-mail: mecanica@ind.uned.es

J. C. García-Prada and C. Castejón (eds.), *New Trends in Educational Activity in the Field of Mechanism and Machine Theory*, Mechanisms and Machine Science 19,
DOI: 10.1007/978-3-319-01836-2_3, © Springer International Publishing Switzerland 2014

1 Introduction

From a political point of view, the interest on distance learning lies on the fact that it constitutes a way to spread education over a wider range of social strata at an economic cost noticeably lower than traditional education. Together with this reason, from a technological point of view, this widespread is possible nowadays more than ever because current ICT allows broad and easy implementations of distance learning methodologies. This assertion could serve by itself to explain the importance acquired by distance education over the last decades.

Alternative strategies to achieve common aims in academic training different from those based on traditional education are possible. This notion is widely defended by many authors nowadays [1], both from the academic and business fields, in such way that even the European Commission gathered this current of thought in one of their White Papers [2] pointing that "*individuals must also be able to enjoy permanent access to a whole range of better targeted and more clearly identifiable education and training provision, which complements general knowledge and can be acquired outside formal systems*". This links immediately with the needed of continuous education for professional training that would be impossible to be accomplished from traditional education systems and results of special interest in the field of engineering.

There is no one single definition of self-education and several conceptions of that idea can be pointed, such as autodidacticism, open education, autocognitive, etc. [3]. What can be stated is that all of them have in common that the learning's initiative is on student's part, choosing pace, circumstances, and in some cases, even the learning strategies.

It can be said that self-education is the basis of distance learning, and it is a fact that the effectiveness of distance teaching programs depend to a great extent on the student's self-education strengthening. However, self-education's ability does not only affect education at a distance but it is also essential in the prospective education of every individual. This is the reason why every education process should have self-education elements and these elements should have also a higher importance as the individual gets older and mature. It is not hard to figure out why many authors [4] assert that self-education must be an objective of formal education in order to achieve an increasing personal autonomy of the student, which would make the individual able to understand new knowledge by himself, giving the people the ability to learn over the rest of his life outside formal systems of education.

2 Distance Learning: Concept and Evolution

The complexity of elements that has contributed to the origin and development on distance learning make very difficult to define the concept of education at a distance due to the different ways the term "at a distance" is understood by the authors, and due also to the high variety of teaching methodologies which exists under this term.

Holmberg [5] states that distance education referrers to the different ways of studying which are not subjected to a classroom continuous supervision by the faculty, but however have planning, guidance and monitoring from a tutorial organization. According to Keegan [6], the distinctive features of this kind of education are the distance between faculty staff and students, the existence of a supporting organization which makes the difference from personal study, the use of technical resources, the existence of a bidirectional communication and a more industrialized set up of education.

Consequently, it can be said that distance learning is an asynchronous learning system, in which the educational action differs in space and time and is characterized by the use of technical resources, the existence of a bidirectional communication between the student and the teacher, a support tutorial organization and an independent and flexible learning system for the student.

It is very common to use the terms of distance learning and open learning as if they were synonyms and more recently, virtual learning and e-learning. However, such terms are far away from being the same as it is stated in the memorandum [7] developed by the Task Force for Human Resources, Education, Training and Youth of the European Commission, which distinguished clearly both concepts

> By Open Learning is meant any form of learning which includes elements of flexibility which make it more accessible to students than courses traditionally provided in centres of education and training. This flexibility arises variously from the content of the course and the way in which it is structured the place of provision the mode medium or timing of its delivery the pace at which the student proceeds the forms of special support available and the types of assessment offered (including credit for experiential learning). Very often the openness is achieved in part at least by the use of new information and communications media.
>
> Distance Learning is defined as any form of study not under the continuous or immediate supervision of tutors but which nevertheless benefits from the planning guidance and tuition of a tutorial organization. Distance learning has a large component of independent or autonomous learning and is therefore heavily dependent on the didactic design of materials which must substitute for the interactivity available between student and teacher. In ordinary face to face instruction the autonomous component is invariably supported by tutoring and counselling systems which ideally are provided at regional/local study centres and to an increasing extent by modern communications media.

Regardless the term referred to, this educational methodology has a long history. For some people, distance learning education history begins with epistolary teaching, going back to the ancient civilizations of the Greeks and the Romans. Remarkable examples through history are those of Pierre de Maricourt, giving explanations about the principles of magnetism by letter to a friend in 1269, or that of *Letters to a German Princess* by Euler in 1770 [8].

Distance learning at a University level appears in America in 1874 with a correspondence program at Illinois Weleyan University, but it was not until 1892 that the first department of distance learning was established by William Rainey Harper, Chancellor of the University of Chicago, under the name of *External Studies Department*. Harper's enthusiasm for the new learning method was such that in 1886 he wrote: *"there will come a day when the correspondence education load will exceed that of our own colleges and schools"* [9].

In the twentieth century this educational methodology has experienced a huge development at all teaching levels, but especially at the university level. Focusing on this field, a decisive milestone is that of the creation, in 1969 in Europe, of the British *Open University* overcoming certain oppositions related to the autonomy and independence of the new university, in order to be equal to rest of the United Kingdom Universities. Since then, several distance learning universities arises successively through Europe, such as the *Universidad Nacional de Educación a Distancia*, UNED, from Spain (1972), the *FernUniversität* from Germany (1974), the *Open Universiteit* from the Netherlands (1982), the *Universidade Aberta* from Portugal (1988). All these institutions are currently integrated in the European Association of Distance Teaching Universities, EADTU.

Towards 1930, near forty North American universities offered distance learning, using the radio besides the postal service. Soon the new technological advances were being incorporated into the learning methodology. Since the 80 years there has been a proliferation in the United States of providers of courses at a distance by TV, via satellite, like those of the National University Consortium or National Technological University, NTU, from Colorado. Also in Canada education at a distance has experienced an important development in the last decades, with examples such as the *Télé-Université by Université du Quebec* (1972), and the Athabasca University of Alberta (1975), among other institutions.

In the Iberoamerican countries there are a large number of institutions with courses at a distance. Distance learning education begins in Mexico in 1972, through *Sistema Universidad Abierta* (SUA) developed by the *Universidad Nacional Autónoma de Méjico* (UNAM) and in Colombia in 1975 with the foundation of *Universidad Abierta de la Sabana* and, in 1982, by governmental decree, the *Consejo Nacional de Educación Abierta y a Distancia*, ordering the *Unidad Universitaria del Sur* (UNISUR) the coordination and total development of the system. In the year 1977 is established the *Universidad Estatal a Distancia* (UNED) in Costa Rica and the *Universidad Nacional Abierta* (UNA) in Venezuela. Cuba also joins this methodology with the creation in 1979 of the *Facultad de Ensañanza Dirigida*, ruled by *Universidad de la Habana*. In Argentina, inside of the National Development Planning 1971–1975, the Ministry of Education included the National Planning of Continuing Education, started by the *Universidad de Buenos Aires*, and in 1983 the *Universidad de Belgrano* created the *Departamento de Educación a Distancia*. In 1980 in Madrid the *Asociación Iberoamericana de Educación Superior a Distancia* (AIESAD) is created encompassing the majority of the Iberoamerican university institutions which use this educational methodology.

In Asia have also arise in the last decades a good number of distance learning university institutions, among which can be cited the Korea Air and Correspondence University (1972), the Everyman's University of Israel (1976), the Sukhtoai Thammathirat Open University of Thailand (1978), the University of the Air of Japan (1981), the Indira Ghandi National Open University in India (1985) and the National

Open University of Taiwan, among others, gathered at the Asian Association of Open Universities (AAOU).

In Africa there exist also examples of distance learning universities, being remarkable that of the University of South Africa which started to teach high-level courses at a distance in 1946. Finally, it has to be said that in Australia distance learning has had a huge and quickly development being the main reason the geographical characteristics of the country: at the early date of 1909 the foundational law of the University of Brisbane, in Queensland, authorized teaching courses at a distance; activity began in the year 1911.

As it can be seen, the development of university teaching at a distance has produced two kinds of institutions: those denominated dual-mode universities, sharing traditional face to face teaching and distance teaching, and single-mode distance universities, using only the latter teaching modality. The experience shows that the number of university institutions that have incorporated this educational methodology to their academic programs has been continuously increasing. The theories that advocate a change from an education based on teaching to other based on learning [10] are reinforcing many ideas which have been the basis of the development of learning at a distance, therefore it would not be a surprise if in the coming years more and more institutions open their doors to this methodology.

Finally, it should be noted that a recient report [11] that compare online and face-to-face learning demonstrates that online learning methods are at least as effective as face-to-face learning.

3 Engineering Education at a Distance

A previous question which has been the center of many discussions among experts and technologist of education is whether anything can be learned at a distance. Empirically, the answer to this question has to be affirmative, since almost every subject and matter can be found in the courses currently offered at a distance. If we accept, as Cirigliano [12], that the representation of face to face versus distance education can be conceived as a continuous line which goes from a total face to face relation between the student and the teacher (classroom-education) to a complete absence of such a relation (self-education), including an intermediate situation of relation at a distance (distance-education), it is possible to understand that any subject can be learned at a distance provided that adequate technological resources are available and there are an enough number of guided practices.

Because engineering at university level needs of a certain percentage of practical teaching, almost every university teaching these topics imposes their students compulsory stays to accomplished laboratory practices. Open University's Summer Schools or lab's compulsory stays at UNED facilities confirm this stance. It is worth mentioning also, that the possibility of successfully implementing high quality distance learning engineering studies has been supported by numerous researchers [13], who explain both the theory and the methodology to do so, and who also describe the many successful previous experiences.

Contrary to general beliefs, the first attempts to teach engineering using distance learning methods can be traced to more than a hundred years ago. As early as in 1903, Julio Cervera founded the "Escuelas Libres de Ingenieros" [14], which was the first attempt in Spain to teach "by mail". In the United States, the University of Wisconsin offered the possibility to study by mail (e.g. by exchanging letters with professors) to the students enrolled in the College of Agriculture that were not able to leave their jobs to attend classes. In Europe, the Open University of Poland opened its doors in Warsow immediately after World War I. This University was aimed at promoting scientific and technical knowledge among all social strata.

Nowadays there are many Universities that offer not only distance learning individual courses, but also full engineering degrees. Table 1 presents some examples of distance learning engineering experiences in Europe. For example, the BSc in Technology at Open University includes degrees in Information Technologies, Pollution Control, Design and Innovation, and Development, Environment and System Practice. Similarly, the *FernUniversität* in Germany offers classes in Electrical engineering and Information Technology as part of the engineering degree. The *Openuniversiteit* in The Netherlands offers classes in Computer Science as part of a degree called *doctoral*, and also to obtain an engineering degree (*ingenieur*). Spanish UNED offers, through the School of engineering, a 5 years engineering degree, with specializations in Mechanics, Electronics, Automatics, and Energy, and offers 4 year degrees in Mechanical Engineering, Electrical Engineering, and Electronical Engineering. In addition, the School of Computer Science at UNED offers 4 year degrees in Computer Science, and in IT Engineering.

Currently more than 17,000 students pursue courses in engineering at UNED. Approximately 2000 are enrolled in mechanical engineering courses [15]. Nowadays

Table 1 Examples of distance learning engineering experiences in Europe

Institution	Country	Teaching	Type
Open University	United Kingdom	Technology	Bsc
		Technology Management	Master
		Mechanical Engineering	
			Specialization
FernUniversität	Germany	Electrical Engineering	Diplom-Ingenieur
		Information Technology	Diplom-Ingenieur
Openuniversiteit	Holland	Computer Engineering	Ingenieur (ir.)
UNED	Spain	Industrial Engineering	Master
		Mechanical Engineering	Bsc
		Electrical Engineering	Bsc
		Electronics Engineering	Bsc
		Computer Engineering	Bsc
		Information Tech. Eng.	Bsc

around ten percent of the students enrolled in the distance learning universities that belong to EADTU, are taking studies conducive to a technical degree.

Additionally, there are some recent platforms such as "Coursera" or "Venture Lab" [16] that offer distance learning engineering classes in universities around the world, although the amount of students taking mechanical engineering classes is still quite small.

There are also some institutions that have courses aimed at offering continuous training to professional engineers. This seems to be one of the areas were distance learning seems to be more useful, as demonstrated by the recent proliferation of this kind of courses in both distance learning universities and traditional universities. As an example, Stanford University has a distance learning program devoted to offer continuous training to companies in Silicon Valley.

4 Methodology

There are differences in the educational methods followed by the different institutions that offer distance learning degrees [17]. There are some similarities in the materials and tools, but there are considerable differences in the emphasis put in each specific tool, and in the way the classes are organized. In the United States, in dual-mode universities, television is widely used. TV programs consist on either programs specifically prepared for distance learning students, or on standard lectures taught to students in a classroom that can be followed on TV by distance learning students. The use of telephone or email, assures that students can ask questions and receive adequate feedback. Some universities have sometimes organized televised conference calls, although its use is not as general yet as the use of TV. In recent years the internet has substituted some of the traditional distance learning methods, providing many more options. European universities, in general, follow similar methods to the ones originally used by the Open University in the United Kingdom. They follow a somewhat more sophisticated pedagogical method than the ones used in the United States, which are on the other hand more pragmatic. In Latin America we can find universities similar to both US and European schools.

In short, the methods used in distance learning teaching are the following:

1. Printed materials
2. Audiovisual materials
3. Tutoring
4. Supporting technologies
5. Evaluation methods.

All the elements of distance learning are managed and organized in the headquarters of the distance learning university, in which professors work, the materials are designed and produced, and the tutoring and evaluation activities are designed and coordinated. Students interact with professors through the supporting technologies (e.g. email, phone). In addition to the headquarters, the distance learning universities rely on an essential part of the system, which are the net of regional centers, in which students can get more personalized attention, and in which the final exams are conducted. This centers receive different names depending on the university (regional centers, study centers, associated centers, etc.), and they differ in the number and scope of services that they offer to students.

In the particular case of engineering studies, all those elements are combined with in-site periods, which are usually short-time visits that students do once or twice a year to conduct lab work, and complete the practical training required by the degree.

5 Conclusions

Distance learning has greatly expanded in recent decades due to socio-economic and political factors, and also due to the development of new information and communication technologies, which has multiplied the interaction possibilities between students and teachers.

The potential of distance learning to train engineers is very large, and has not been yet exploited adequately. Neither the traditional nor the distance learning schools have yet sufficiently exploited the possibilities of distance learning in two of the most demanded areas: college degrees in engineering and continuous training of engineering professionals. The flexibility and openness of distance learning makes it particularly adequate for this latter type of education.

Finally, the recent appearance of global distance learning platforms, and the increasing offering of continuous training courses in other areas, anticipates that in the next few years, distance learning courses in mechanical engineering as well as other technical fields will experience an exponential growth.

References

1. Artés, M.: "Formación a distancia de ingenieros", en Formación de ingenieros. Objetivos, métodos y estrategias, pp. 391–405. ICE UPM, Madrid (2005)
2. CCE: Livre blanc sur l'education et la formation. Enseigner et apprendre. Vers la societé cognitive. EPOCE, Luxemburgo (1995)
3. Carré et al.: L'autoformation. Presses Universitaires de France, Paris (1997)
4. Sarramona, J.: La autoformación en una sociedad cognitiva. Revista Iberoamericana de Educación a Distancia, vol. 2, nº 1, junio, pág. 41–59 (1999)
5. Holmberg, B.: On the Status of Distance Eeducation in the World in the 1980s. ZIFF, Hagen (1985)
6. Keegan, D.: The Foundation of Distance Education. Croom Helm, Londres (1986)
7. CCE: La enseñanza abierta y a distancia en la Comunidad Europea. Luxemburgo (1991)
8. Baat, J.A.: Postal two-way communication in correspondence education. Liberhermods, Malmoe (1985)
9. Mackenzie, O., Christiensen, E.L.: The Changing World of Correspondency Study. Pensilvania Sate University Press, College Park (1971)

10. Chickering, A.W., Gamson, Z.F.: New Directions for Teaching and Learning: Applying the seven Principles of Good Practice in Undergraduate Education. Jossy-Bass, San Francisco (1991)
11. Means, B., Toyama, Y., et al.: Evaluation of Evidence-Based Practices in Online Learning: A Meta-Analysis and Review of Online Learning Studies. U.S. Department of Education, Office of Planning, Evaluation, and Policy, Development, Washington (2010)
12. Cirigliano, G.F.J.: La educación abierta. El Ateneo, Buenos Aires (1983)
13. Villarroel, A., y Pereira, F. (eds.): La educación a distancia: desarrollo y apertura. ICDE, Caracas (1990)
14. Sarramona, J.: La enseñanza a distancia. Posibilidades y desarrollo actual. CEAC, Barcelona (1975)
15. UNED: Memoria anual. http://portal.uned.es/portal/page?_pageid=93,25170900,93_25170901&_dad=portal&_schema=PORTAL (2010)
16. Reyes, H.: Cronología de la revolución de la educación 2012, desde OpenCourseWare y Khan hasta Coursera. WeduboX y Udacity, Educación Virtual (2012)
17. García Areitio, L.: Educación a distancia hoy. UNED, Madrid (1994)

Basic Projects of Knowledge Integration in the Bachelors' Degree in Industrial Technology

S. Cardona Foix, L. Jordi Nebot, R. Pàmies-Vilà and P. Català Calderon

Abstract The syllabus of bachelors' degree in Industrial Technology at Barcelona School of Industrial Engineering has Project I and Project II on semesters 4 and 6. In these subjects, students must face correctly to a concrete engineering problem which has to structure and solve with the knowledge and competences acquired in previous subjects. In the frame of Mechanical Engineering, it is recommended that the problem to solve by the students should involve the realization of a task within a technological process that requires the manipulation of a real mechanism. The project includes the identification and concretisation of the task and the technological process, the geometric and kinematic analysis of the mechanism, the design of a manoeuvre, the implementation of the control of the actuators for the execution of the manoeuvre and the design and construction, with rapid prototyping, of an adequate accessory for the task. Till nowadays, the results after three semesters of teaching have been satisfactory from the teachers' and students' side.

Keywords Projects · Integration · Knowledge · Competences

S. C. Foix · L. J. Nebot (✉) · R. Pàmies-Vilà · P. C. Calderon
Department of Mechanical Engineering, ETSEIB, Universitat Politècnica de Catalunya, Catalunya, Spain
e-mail: lluisa.jordi@upc.edu

S. C. Foix
e-mail: salvador.cardona@upc.edu

R. Pàmies-Vilà
e-mail: rosa.pamies@upc.edu

P. C. Calderon
e-mail: pau.catala@upc.edu

J. C. García-Prada and C. Castejón (eds.), *New Trends in Educational Activity in the Field of Mechanism and Machine Theory*, Mechanisms and Machine Science 19,
DOI: 10.1007/978-3-319-01836-2_4,

1 Introduction

The syllabus of the Bachelor's Degree in Industrial Technology (GETI), at Barcelona School of Industrial Engineering (ETSEIB) includes the subjects Project I and Project II in semesters 4 and 6—S4 and S6 [1]. These courses are designed for promoting, encouraging and empower the vision of engineering science and technology in the early stages of the training of students. Therefore, in these subjects, the students must correctly pose a particular engineering problem and solve it concretely using the knowledge and skills acquired in previous subjects or in subjects that are pursuing in parallel with Project I (this is the case of the Theory of Machines and Mechanisms, for example).

The first courses of GETI have a strong bias towards the core subjects, mathematics and basic sciences, and the technology is reduced to a marginal situation. This fact contrasts with the reality of the students who have chosen this degree at ETSEIB as a first option after passing the university entrance exams (PAU). Note that, the first option might have been for more specific training centres in basic sciences, but they have not chosen this option. Some initial bias towards training in mathematics and basic sciences is accepted and justifiable because the students have diverse origins and they arrive at school with different levels on basic topics. Achieving competence in core subjects at the university level must be ensured. However, in previous curriculum, the existence of this bias caused some students to drop out the school and maybe these students could have been totally appropriated for engineering fields.

The inclusion of Project I and II in the syllabus and the proposed approach, attempt to combining, in the initial courses, the need to provide the appropriate level in core subjects with the opportunity for students to use their skills and applying their knowledge to technological challenges.

During the implementation of the syllabus, the board of direction of the ETSEIB asked to the departments to propose engineering problems that might be studied in Project I and II. The authors of this article, who also teach Theory of Machines and Mechanisms and Mechanical Vibrations, decided to participate in this stimulating demand, using the background and experience acquired previously in two areas: teaching the mentioned subjects and also organizing, defining and preparing the university entrance exams of the technology area.

Authors' experience with PAU exams dates back to 1999–2000. Once studied and analyzed the curricula of the subjects in the area of technology, also the documents of the Accreditation Board for Engineering and Technology [2], and the Higher Engineering Education for Europe, among others, the authors have suggested some exams. The main goal of these tests is to evaluate the knowledge and the abilities acquired during upper secondary school education. The exams are prepared in the way that students with a vocation and training in technology could successfully overcome and feeling this procedure as an introduction to engineering.

The authors also developed several documents that summarize the knowledge that students who want to pursue an engineering degree should have, regardless of the legislative framework [3]. The ideas reflected in these documents apply to all students whether or not any direct contact with technology in high school or in previous studies.

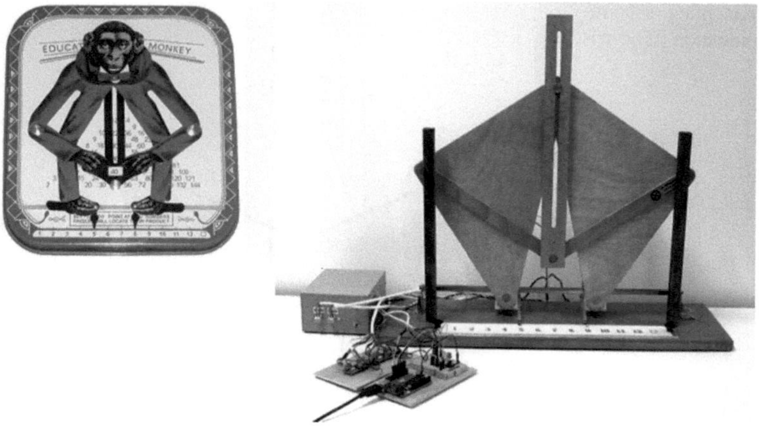

Fig. 1 Automaton "The wise monkey" and its operational model

2 Approach

This section reflects the main ideas exposed in the cited documents applied in the environment of GETI and specifically in the field of Mechanical Engineering [4]. In order to achieve the objectives of Project I and II, the students have to perform a task, if it is possible related within a technological procedure, which requires the use of a machine or mechanism with operational models are available.

Students have used different mechanisms along these years: a pantograph parallel manipulator, a straight line manipulator or an automaton (Fig. 1). The project includes the identification of the specific tasks to develop, the technological process requirements, the geometric and kinematic analysis of the mechanism, the design of a manoeuvre, the implementation of the actuators to control the execution of the manoeuvre and the design and construction, through rapid prototyping, of a tool or a suitable accessory for the mechanism proposed task.

The descriptions will focus, where appropriate, on the automaton shown in Fig. 1, whose standard symbol kinematic scheme is shown in Fig. 2. A prototype of this mechanism is available in the Laboratory of Machines. The prototype has been designed developed and built in the own laboratory. All student teams have worked with the same model. On one hand, asking each team of students to design and build their own model require knowledge and skills that students do not necessary have. Moreover, it represents a high economic cost and the use of workshop hours (and resources) which really are not assumable. On the other hand, the use of standard elements (Meccano®, Fisher®…) would solve the problem of workshop hours but limits the design freedom. At this time, this option has not been contemplated. Furthermore, the skills required for the mechanical design of the model are not intended to be acquired in undergraduate studies at ETSEIB.

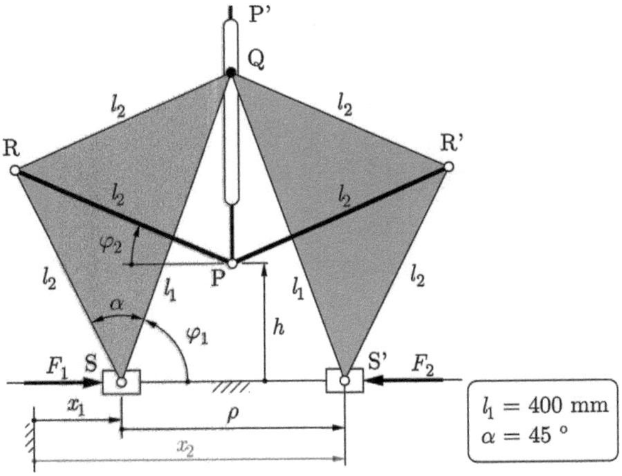

Fig. 2 Kinematic scheme of the automaton "The wise monkey"

The approach of this project has forced teachers remembering and conveying students the need to have an overview of all the fields involved in the operation of a machine. This vision is sometimes lost because of focusing too much on the same subject or a topic of research, and perhaps our vision is far from the reality in the production engineering.

The authors, teachers of Theory of Machines and Mechanisms and Mechanical Vibrations, have had to update their knowledge in the areas of programming, electronics and control and the use of existing tools.

It should be noted that the use of sensors for feedback and control of the movement is not envisaged, since it is an unknown knowledge for students of S4.

It is also considered that while students are able to perform some relevant dynamic calculations, is better not using them in Project I. Without forgetting to make students be aware that it is a limitation of the project, the authors not include this study in favour of reaching a tangible result on the available time. Thinking of an incremental learning process, the dynamic aspects can be used in Project II.

3 Methodology

The 12 lessons of 2 h are taught in a classroom using a Tiddly Wiki application. This application contains all the necessary information for the development of the project and is organized by topics, some of which include exercises that students must complete to become familiar with the more complex aspects of the project.

In all sessions, the teacher explains the tasks that students have to develop before the next session. All these tasks drive students until the realization of the final project. Furthermore, this procedure is designed in order to achieve the generic skills associated with the own project—effective oral and written communication, self-learning and teamwork.

The basic tasks to be performed by the students are:

- Find information about the evolution of automata over the years and in different cultures.
- Perform a kinematic analysis of the automaton.
- Design an appropriate mechanism (if they propose any variant to assemble in the original one).
- Implementation of the prototype control with a microcontroller.
- Execution of the graphical representation and the drawings of the mechanism.
- Design of a tool or accessory that could be materialized using a 3D printer.

The organization and distribution of team tasks is left up to each group of students to enable them implementing the concepts that are explained in the lectures of the course (3 sessions taught by the Department Engineering Design). These lessons are often associated with big companies' large projects—departmental divisions, severely division of tasks, responsibilities...—or with working methods that are usually applied in small and medium enterprises, technical offices of free exercise of the profession... Teachers only suggest to avoid duplication of efforts. The proposed dedication outside the classroom of each student is about 3 h per session, and therefore about 36 h per student. The expected work is evaluated for 36 h multiplied by the number of team mates (between 3 and 5). As known, one student can work more intensively to one part of the project. However, everyone should know the whole project and be able to defend a reasonable answer to any questions about it. Finally, it should be clear what the exact contribution of each member of the team is, in the whole project.

4 Evaluation

The evaluation system is common to all groups on Project I and II. The final mark—N_{final}—is obtained as:

$$N_{final} = 0,15\, N_{p1} + 0,2\, N_{p2} + 0,4\, N_f + 0,25\, N_{ind}$$

N_{p1}, N_{p2} and N_f are team marks and they evaluate written reports and oral presentations; N_{ind} is an individual mark taking into account the continuous assessment of the student.

Rubrics are used to evaluate the oral and written presentations. These rubrics evaluate specific aspects as either "very good" (a score of 8–10); "satisfactory" (a score of 5–7); "insufficient" (a score of 4–3) and "totally unsatisfactory" (a score of 2–0).

The rubric for evaluating written work assesses the graphical representation and the plans of the mechanism, the written presentation of the work, the presented objectives, the obtained results, the discussion of the key issues, the deduced conclusions and the correctness in the reference list. The rubric used during the oral presentations is focused on evaluating the content, the order and organization of the presentation, the verbal and nonverbal skills of students, the graphical resources used and on the ability to respond the questions asked by teachers.

5 Material Used

The model of the automaton used is made by wooden links with metal joints. This structure makes it very lightweight and safe handling. The use of this type of construction can be attractive to a large group of students who can directly manipulate, for the first time, a little machine. Furthermore, they have a reference on how to use it, since it is also the type of construction used in a large number of models presented in Theory of Machines and Mechanisms (both, in theoretical classes and also in laboratory sessions). These models are used often as an intermediate step between the reality (shown directly or through videos or photographs) and the standard symbols mechanism kinematic schemes.

In order to focus the points of interest and to raise issues affordable for students, the drive controller is done using stepper motors. These motors allow students to avoid the problem of feedback. However, a strategy to define the origin of the movement of the motors is needed. Its measurement and load capacity are reasonable and the fact that they are widely used makes them accessible both for its location and its economic cost (Fig. 3).

The motor control signal and auxiliary signal elements, the sensor readings, the necessary calculations for generating the response of the system from inputs, and the communication with the computer are performed by an electronic board *Arduino*. This is an open electronic platform to create prototypes based on flexible software and hardware. It is an easy tool to use in a simple environment. The board has a

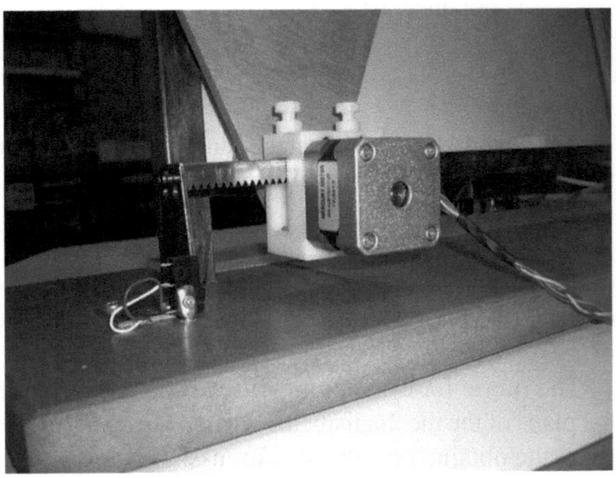

Fig. 3 Stepper motor and assembly in the model

Fig. 4 3D printer that work by adding material, and detail of its extruder´

microprocessor Atmel and is programmed in C code. The program is loaded into the microprocessor, via USB bus, of the plate and it can operate autonomously. From the website of the company [5], the integrated development environment (IDE) can be downloaded. This application can be used for programming and loading the program into the microcontroller.

A prototyping board, a button, a LED array, cables and resistors are provided to students in order to become familiar with the programming of the microcontroller and with the use of the Arduino system. They have to design a system allowing turning on and off the LEDs following some orders that correspond to a realistic application: simulating car semaphores, combining car and pedestrian semaphores, highway toll stations… Interestingly, they develop and explore new situations that are not the typical examples that can be found, for example, in the Internet.

The required calculus are made using SciLab®. Scilab®is a free software for numerical calculation, open source and multi-platform. The programming and operation is similar to MatLab®. Since it is free, students can use it on any computer, allowing them to perform their meetings inside or outside the university.

Some templates for programming tasks (calculation and simulation of the mechanism and also control system implementation) are provided to the students. These templates have to be modified and adapted to the own application of each team.

Finally, and in order to manufacture the components for the application that each team of students has decided, a rapid prototyping machine is available in the laboratory. Figure 4 shows the machine which creates objects by adding material techniques.

6 Results

Till now, all teams have successfully completed all the stages of the project. It is important to remark the achievement of the precise definition of realistic tasks and their correct implementation. Thus, for example, although the original version of the automaton shows the result of multiplying two integers between 1 and 12, some students have educated the wise monkey to calculate the average distance between the planets of the solar system, or to obtain the result of adding two numbers. All teams have managed to define a realistic and specific task for each stage of the project.

The evolution of the written reports from the first submission (during the third session) until the final report has been spectacular. The first draft deliver is usually no more than a set of information collected from the Internet without any structure. A conceptual and formal elaboration can be appraised in the last report. Generally, it is a totally acceptable technical report. The same behaviour can be observed for the oral presentations.

7 Conclusions

The tasks have been made responsible, in most cases, within schedule and with satisfactory results.

All the students, who have participated in the projects till now, have shown a great satisfaction when the system finally performs what they have initially in mind. At the beginning of the course, they were really pessimistic and thought that the tasks represents a lot of work and dedication. However, when they achieve the final result, they enjoy their own work.

The students also recognized to be satisfied with the approach of the project. It represents, for many students, the first opportunity during the degree to combine the knowledge acquired at different previous subjects. Moreover, the project provides a first overview of a whole procedure, something that is necessary working as an engineer.

The teachers responsible for the project, while acknowledging that the dedication to them is really considerable, are fully satisfied with the results. They believe that the approach is appropriate and they think that new topics can be proposed using the same methodology. This implies a major renovation and continuous effort to develop new mechanisms and novel projects.

The teachers believe that it is worth making this effort: the students who take this type of project are very happy and they have a better relationship with the Theory of Machines and Mechanisms environment than students who only attend regular courses.

References

1. http://www.etseib.upc.edu/ca/cursactual/portal-dassignatures (2013). Accessed 12 March 2013
2. http://www.abet.org/accreditation-criteria-policies-documents/ (2013). Accessed 12 March 2013
3. Materia de Tecnologia Industrial per a les PAU (12 March 2013)
4. Singhose, W., Donnell, J.: Introductory Mechanical Dessign Tools. Georgia Institute of Technology, Atlanta (2009)
5. http://www.arduino.cc (2013). Accessed 12 March 2013

Content and Realization of Education in Mechanism Theory at RWTH Aachen University

S. Kurtenbach, T. Mannheim, M. Hüsing and B. Corves

Abstract Within the last years new trends in education in mechanism theory have been established. For improving the sustainability of learning the conventional separation into lectures and exercises is extended by a laboratory which is held for a small audience. Because of this it is possible to enhance the education in mechanism theory by using the computational methods within the process of understanding. The students are familiarized with those methods, e.g. the geometry software Cinderella, in a very early stage of mechanism development. Especially different synthesis and analysis procedures are applied in this stage. At RWTH Aachen the courses "Electromechanical Motion Technology", "Motion technology" and "Kinematic, dynamic and applications in robotics" are taught in Bachelor and Master studies. In this chapter we will present the implementation of the innovative educational concepts basing on the above mentioned trends. Firstly, we will introduce the content of these courses. Therefore, the reason for segmenting the content into the different classes will be explained. Subsequently, additional background information (group size, lecturers, coordination between lecture, exercise and laboratory) are given and the complete organization of the courses is explained as well. The concept is demonstrated applying a vivid and descriptive example.

Keywords New media for teaching · Internet learning · Novel educational concepts

S. Kurtenbach (✉) · T. Mannheim · M. Hüsing · B. Corves
Department of Mechanism Theory and Dynamics of Machines, RWTH Aachen University, Aachen, Germany
e-mail: kurtenbach@igm.rwth-aachen.de

T. Mannheim
e-mail: mannheim@igm.rwth-aachen.de

M. Hüsing
e-mail: huesing@igm.rwth-aachen.de

B. Corves
e-mail: corves@igm.rwth-aachen.de

J. C. García-Prada and C. Castejón (eds.), *New Trends in Educational Activity in the Field of Mechanism and Machine Theory*, Mechanisms and Machine Science 19,
DOI: 10.1007/978-3-319-01836-2_5,

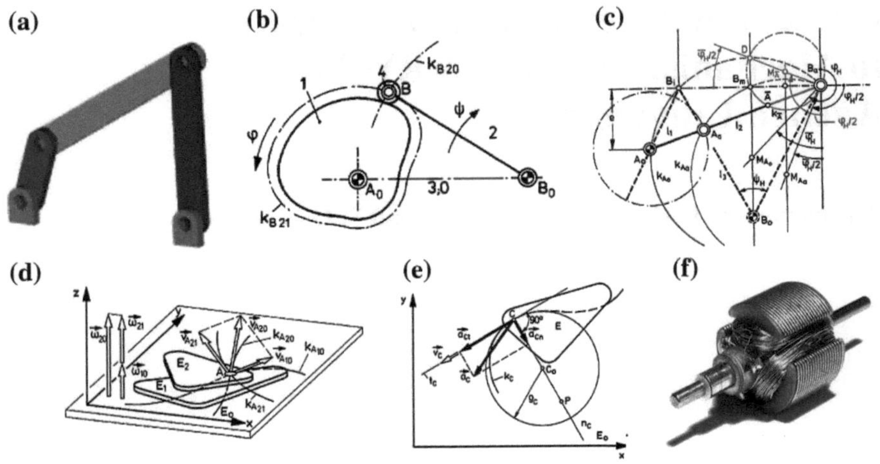

Fig. 1 Principles taught in the course EMAT

1 Introduction of Courses

This chapter introduces the content of the several courses whose framework and innovative education trends were presented in this chapter. In particular, the different kinds of lectures, exercise lectures and laboratory courses were explained [1].

Within the field of Mechanism Theory at RWTH Aachen University the department of Mechanism Theory and Dynamics of Machines (IGM) offers in total three different lectures. These are "Electromechanical Motion Technology", "Motion technology" and "Kinematic, dynamic and applications in robotics". The detailed contents are introduced in the following.

The lecture "Electromechanical Motion Technology" (EMAT) deals with the determination of the kinematic parameters exclusively of planar mechanisms which can be transfer and guidance mechanisms. In particular crank (Fig. 1a) and cam mechanisms (Fig. 1b) as well as their hybrids were introduced. Regarding the crank mechanisms one focus is the synthesis of these mechanisms in both ways, structural synthesis and dimensional synthesis. Basic principles like the position synthesis and dead center position synthesis (Fig. 1c) are taught. Another important issue is the analysis procedures for determining the position, velocity (Fig. 1d) and acceleration (Fig. 1e) state of a certain link of a joint. The course ends with presenting different types of electric drives (Fig. 1f). Especially the rotational and electric drives are focused to basically get introduce drives for planar mechanisms. Therefore many examples from very different areas in the field of engineering were analyzed. The students were faced to this required course in the second half of the Bachelor studies of mechanical engineering. More exactly it is in the fifth semester of the Bachelor studies where they have to firstly specialize in their degree program.

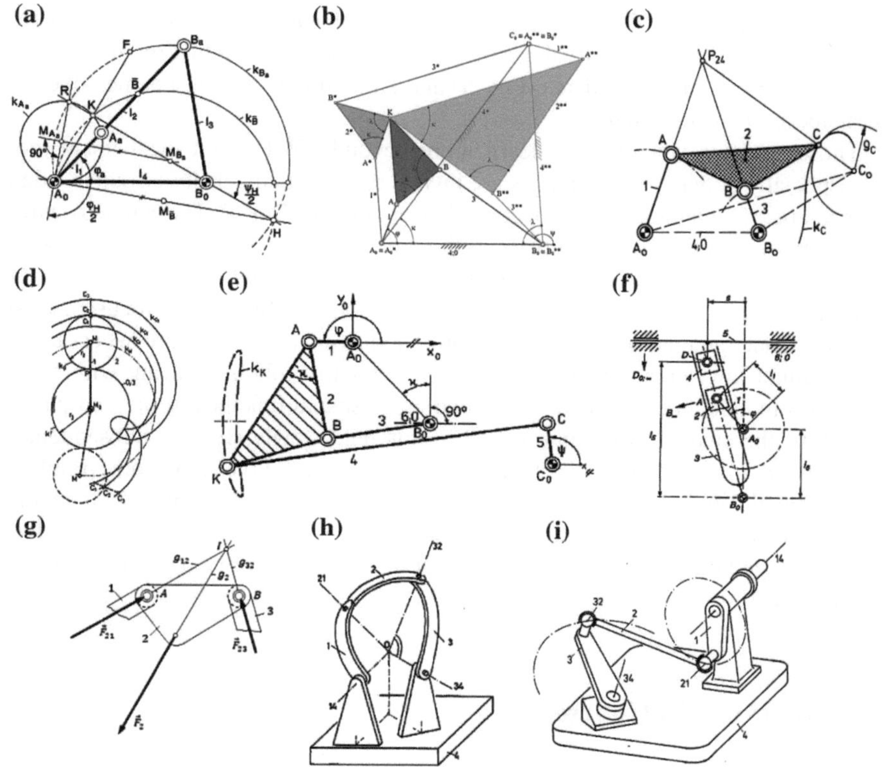

Fig. 2 Principles taught in the course BWT

Just the basic foundations for dimensioning a mechanism were taught in the course EMAT. Students doing their master get in touch with a second required course offered by the IGM in the second semester of the Master studies. This course is called "Motion Technology" (BWT) and the students were confronted with highly advanced problems and solving procedures within the field of mechanism theory. In this course planar linkages with up to six links were considered with all the techniques explained in the following. Mechanism analysis procedures for five and six bar linkages and classification schemes for motion tasks were presented. Additionally advanced procedures for the dead center position synthesis (Alt's theorem) (Fig. 2a) were introduced as well as a technique to multiple generate coupler curve (Roberts Mechanisms) (Fig. 2b). The determination of centrode tangent and normal and subsequent application of bobillier's theorem (Fig. 2c) and the equation of Euler-Savary is another new method for the students. Furthermore the students are confronted with new types of mechanisms. There are the wheel (Fig. 2d), dwell (Fig. 2e) and synchronous mechanisms (Fig. 2f). Wheel mechanisms in planar, spherical and spatial configuration were introduced. There, the different types of wheel gears (one fixed wheel, wheel-crank-mechanisms, sumgears, gears with very high transmission ratio) were introduced. The last topic regarding the planar mechanisms is the kineto-static analysis (Fig. 2g). Finally the course ends up with the introduction of spherical (Fig. 2h) and spatial mechanisms (Fig. 2i) and all their calculation schemes like the Hartenberg-Denavit-Notation.

Those students who want to learn more beyond BWT about mechanism theory especially about spatial linkages are invited to listen to an elective course called "Kinematics, Dynamics and Applications in Robotics" (KDaRo). The course KDaRo gives a brief overview on main parts in the field of spatial linkages and in general robotics. The mechanical design as well as the kinematic and dynamic analysis of serial and parallel robots, which are a special case of spatial linkages, is focused. Goal is to introduce the methods of the kinematics and dynamics, which are necessary to select, dimension, move, control, program and control robots. Starting with basic principles to calculate the degrees of freedom of spatial linkages, serial and parallel robots, the students were introduced in this field slowly (Fig. 3). Subsequently vectorial calculation procedures to solve the inverse and direct kinematic problem are presented. Within the framework of a structural synthesis the optimal kinematic structure can be selected using a classification for robotic structures. The Euler-Newton procedure and a recursive algorithm to solve the inverse dynamic problem are derived.

2 Formation of a Course

All the courses in Mechanism Theory at IGM are divided into the three parts, lectures, exercises and small-group-exercises. The student are faced with the theory of each topic for the first time in the lecture. This lecture is held by the professor of the department, who is teaching the theory by deriving it from before known knowledge. Hence some topics are mainly graphically the professor is using a geometry software called Cinderella to visualize the graphical construction as well as to proof the correctness of the constructions by moving the mechanism and showing the correctness of the derived theory [2].

The exercise is held by a lecturer. In the exercise following the lecture the summary of the theory is explained again by the lecturer. The application of the theory will be taught by solving problems of the practice by the students. The teaching is made more efficient by giving the students time to solve the examples on their own, if they are stucked they can get some help. Afterwards the correct solution will be presented. This approach makes the students think about the problems on their own before getting the solution to know and helps to understand the solution.

In subsequent arranged and by the students optionally selectable small-group-exercises the knowledge from the previous mentioned courses can be engrossed in the mind. This course is executed by student coworkers who already passed the corresponding exam. At the beginning there is a small input of theory repeating the subject matter of both, the lecture and the exercise. The topics within this small-group-exercises don't differ from both previously held courses. It is simply another

Fig. 3 Principles taught in the course KDaRo

philosophy of education where the tasks are introduced at the beginning of each lesson and the students need to solve it on their own. The student coworkers of course guide clueless students to the solution but, and that is a huge advantage, without telling the solution itself. In a small discussion about the task the problem nearly is solved doing nothing but a conversation about the theory. Usually this course starts with a solution of the task on a blank piece of chapter implementing the graphical solution procedure learned theoretically in the lecture and in the exercise lecture. Once the solution for the particular position of the mechanism given on the sheet is found, the overall solution for an arbitrary position subsequently is implemented using the geometry software Cinderella interactively.

Some students being eager of knowledge to hear more about the practical implementation of graphical synthesis and analysis procedures in the field of mechanism theory there is offered another additional voluntary course which is called "Computer-Aided synthesis and analysis of mechanisms". Within this block course the students no longer use simple planar geometry software but extensive CAD-Software like Autodesk Inventor. Here the graphical synthesis and analysis procedures of EMAT and BWT are repeated in the sketch-environment of the CAD tool like in [3, 4]. Furthermore they are executed in the assembly-file and new topics like a spherical position synthesis procedure are introduced. Huge advantage of this approach is the expertise knowledge the student gets by using the software. They interactively experience that all the theoretically mentioned procedures from the lec-

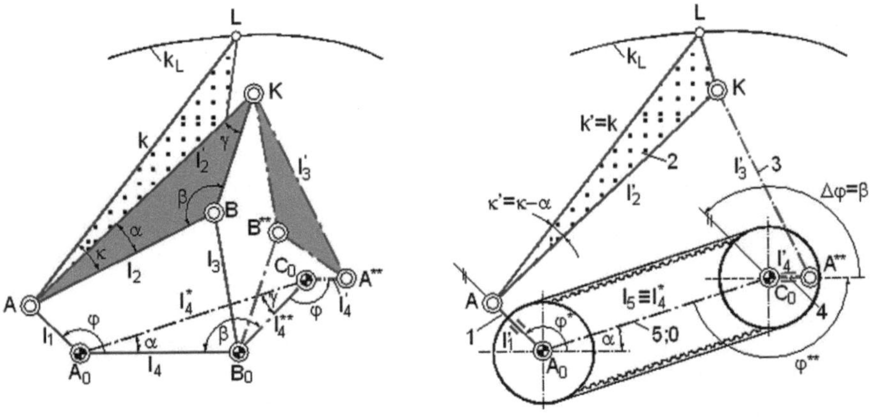

Fig. 4 Construction of roberts five-bar-mechanism

ture are directly implementable without problems in a CAD-Software. Later on, the young engineer in industry knows how to apply these procedures even when there is no simple geometry software available.

3 Examples

The before described segmentation into lecture, exercise and small group exercise will be shown by the example of the Roberts Mechanism (Fig. 4). The theory of the multiple creation of coupler curves with the Roberts Mechanisms is taught in the lecture. The construction of the three possible four-bar-mechanisms is shown in theory. The possibility of extending the Roberts-Mechanisms towards a six-bar-linkage a five-bar-mechanism with an arbitrary fixed-point is taught.

In the related exercise the way of constructing the Roberts-Mechanisms is shown. The students are constructing the mechanisms by doing it for practical examples. An example for a task is shown in Fig. 5, here a film transportation mechanism is shown. The fixed joint B_0 is in the space of the film cassette, another mechanism performing the same coupler curve is needed to construct the transportation mechanism. The limitation of design space shows a common problem in practical applications of mechanisms, the students learn using the theory here. In this example there is no four-bar-mechanism with both fixed joints in the design space, thus a five-bar-linkage is required. This example is solved with pen and chapter. It is impossible to see the correctness and function of this construction, thus the students can visit the small group exercises to do a hands-on-analysis. Using the geometry software Cinderella it is possible to do the construction again and let it run afterwards. Within the small group exercise the student coworkers are explaining the construction again in a very small theory block at the beginning. In this exercise segment the focus is on the

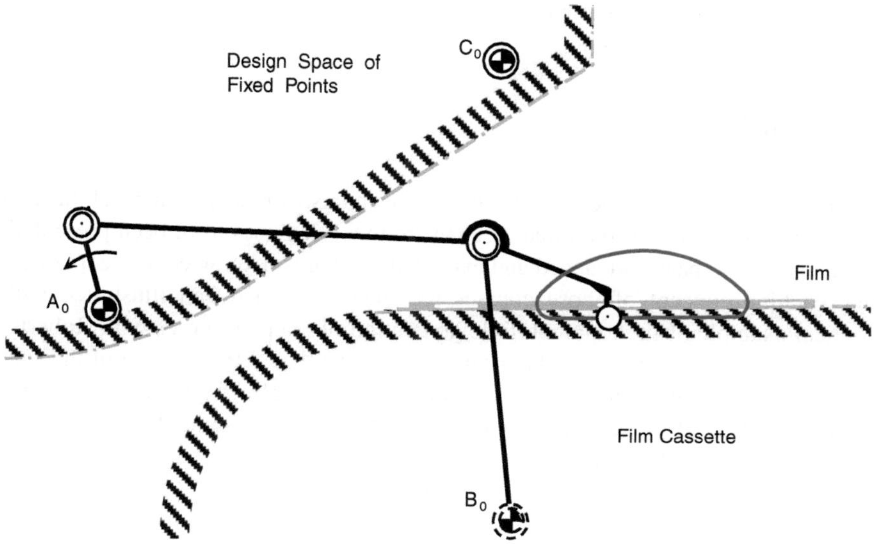

Fig. 5 Photo film transportation mechanism [5]

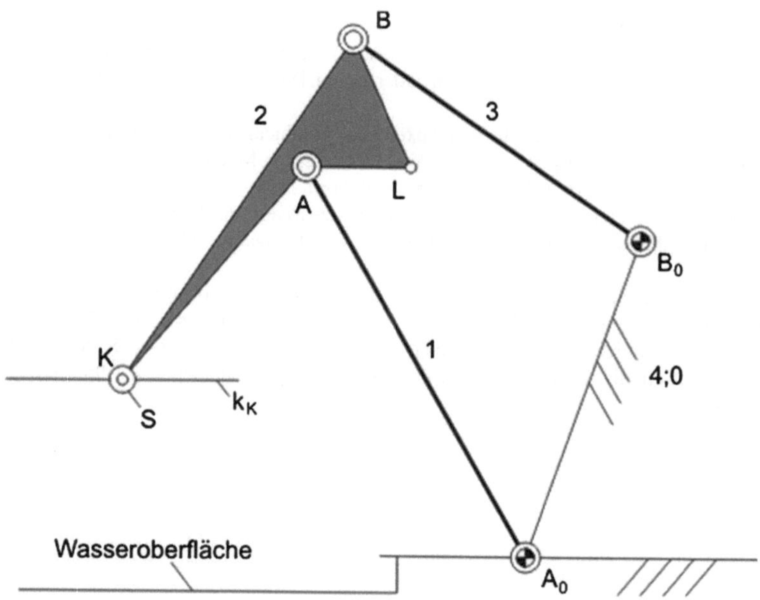

Fig. 6 Schematic view of a port crane

design of mechanisms with the help of the computer. Thus the students will have some time to do the examples of the exercise with help of the computer and will get some new examples, like shown in Fig. 6, to solve.

4 Conclusions

Within this chapter the teaching concept of mechanisms theory at RWTH Aachen is shown. The content of the three mechanism theory courses have been presented. The division into basic, advanced and extended mechanism knowledge is explained. The teaching concept of those courses has been explained and illustrated with practical example of the topic multiple generation of coupler curves with Roberts-Mechanisms. The repetition of theory in different depth and the concept of the segmentation have been evaluated. The students gave best marks and are using this offer extensively. The effort leads also to improved marks in the exams. Because of this we think this concept is really successful.

References

1. Barej, M., Hüsing, M., Corves, B.: Teaching in mechanism theory—from hands-on analysis to virtual modeling. In: New Trends in Mechanism and Machine Science; 4th European Conference on Mechanism Science, pp. 703–710. Santander, Spain. ISBN: 978-94-007-4901-6, ISSN: 2211-0984, 18-22 Sept 2012
2. Corves, B.: Computer-aided lectures and exercises: graphical analysis and synthesis in mechanism theory. In: Proceedings of the 11th World Congress in Mechanism and Machine Science, vol. 2. China Machine Press, Tianjin, China, 1–4 April 2004
3. Lonij, G., Kurtenbach, S., Corves, B.: Parameter-driven mechanism synthesis in CAD. In: International Design Conference: Design 2012, Croatia, Dubrovnik. ISSN: 1848 4700, 21–24 May 2012
4. Lonij, G., Kurtenbach, S., Hüsing, M., Corves, B.: Interactive design of opening and closing mechanisms for skylight domes. In: New Trends in Mechanism and Machine Science, pp. 227–237. Springer, Dordrecht Heidelberg. ISBN: 978-94-007-4901-6, ISSN: 2211-0984
5. Kerle, H., Pittschellis, R., Corves, B.: Einführung in die Getriebelehre, p. 195. ISBN: 978-3-8351-0070-1

Integration Processes in Engineering Education

V. V. Kuzlyakina

Abstract Integration processes cover more and more spheres of activity of the human community. Joining of many countries to Bologna's convention on education and development of up-to-date computer technologies open new possibilities for integration in engineering education too. Discipline Theory of Mechanisms and Machines is fundamental in training of any engineering specialists. The objective scientific laws of mechanics being in the basis of this discipline and other engineering courses are general for the world community. Scientific, methodical and technological provision of course TMM varies in different countries. One country is remarkable for its high scientific achievements, another—for methodical discoveries, the third country is good at experimental provision. It is reasonable to unite the best achievements for making a computer and mechanical complex on engineering disciplines. This can serve the base for integration of scientific, methodical and technological potential for creation modern education facilities applicable in different countries. There is a necessity now to create modern complexes integrating mechanical experimental stands with up-to-date measuring means, computer processing of experiment results and a remote access for providing the technology of a distant education.

Keywords Integration processes · Computer and mechanical packages · Computer processing of experiment

V. V. Kuzlyakina (✉)
Maritime State University named after adm. G. I. Nevelskoy, Vladivostok, Russia
e-mail: kuzlyakina@msun.ru

J. C. García-Prada and C. Castejón (eds.), *New Trends in Educational Activity in the Field of Mechanism and Machine Theory*, Mechanisms and Machine Science 19,
DOI: 10.1007/978-3-319-01836-2_6,

1 Introduction

Theory of Mechanisms and Machines is fundamental in training of any engineering specialists. The objective scientific laws of mechanics being in the basis of this discipline are general for the world community and can serve the base for regional and international integration in engineering education.

Mechanics, as well as many sciences began to burgeon in the seventeenth century. A large amount of applied engineering sciences based on classical mechanics, which developed rapidly, using more sophisticated mathematical tools, modern technology and computing facilities burgeoned in the twentieth century. There was a situation when general professional disciplines began more intimate and unrelated. This circumstance did not allow integrating them for solution of complicated problems of design and for operation of specific objects. At the same time the educational program for training of engineering specialists to decrease the amount of class hours on general professional disciplines. Sometimes such an important subject as the theory of mechanisms and machines was displaced; amount of calculation and graphic works was decreased (from course project to course paper, then to calculation and graphic work and then there is the only pass–fail exam).

Development of computer training technologies (CTT) allows carrying out the integration of engineering disciplines up-to-date [1]. CTT is a set of facilities; modern tools of training organization and creation of the information environment; professional software complexes for solution of applied problems; highly qualified specialists who are in possession of modern tools of engineers and teachers. The main functions of CTT in technical university are development of creative engineering thinking, training of student to use the computer and design engineering packages which are required in future productive activity.

2 Development of Technology and Science of Machines

Scientific-and-technological progress develops because of three main reasons: wars (unfortunately), laziness (mechanization of technological procedures) and love (of work, of country, for family). Periods of technology development are described shortly bellow.

In the first century Vitruvius described a lifting device of garrison gates, mills mechanisms; mechanical clock appeared in the tenth century; printing press—in the eleventh century; a tremendous upgrowth of technology in Italy in the thirteenth century, the great Leonardo da Vinci designed a number of different mechanisms in fifteenth century. In Russia in the seventeenth century there were large complex mechanisms: mechanical bellows; forging hammer; reciprocating saw; big drills for fabrication of main guns activating by hydro-power. In the end of the seventeenth century there was steam engine for water pump drive in France. The eighteenth century is a century of steam engine: Newcomen created a steam engine in 1711;

Table 1 Periods of technology development

Period	≈Class number of engineering systems	Average number of elements in engineering systems
100,000 years ago	5	1
10,000 years ago	50	10
1,000 years ago	1,000	100
Nowadays	50,000	10,000

regardless of Newcomen in 1763 Polzunov I. I. created a steam engine, in 1765—Watt, a patent on double-acting steam engine was received by Watt in 1782.

Periods of technology development are represented in Table 1. Development of science of machines began early in the nineteenth century. It was a century of empirical mechanical engineering.

The first course on TMM was 1808 in the Ecole Polytechnique and in 1811 in Russia in the Petersburg Institute of the Transport Engineers Corps. The first textbook on TMM in Russia was written in 1823 by professor Chizhov N. A. at St. Petersburg University. The most tremendous upgrowth of science of machines was in the first half of the twentieth century. At this period applied engineering sciences had a tremendous upgrowth and became unrelated. Development of analytical methods of research and occurring of super-computing complexes were helpful for integration of applied engineering disciplines. In the nineteenth century the inventor was an engineer, a technician, a draftsman and a manufacturer of his invention. He kept in mind all the parameters providing their conformance.

While complication of technical system (Table 1) the number of people providing manufacturing of technical object increased, there were different structures: technology and design bureau, service of preproduction and control of production and others. It became more complicated to provide the relationship between parameters. There was a necessity for a complex automation of production, technical operation and disposition of technical object [2]. The structure of a complex production system, technical operation and disposition of technical object is represented on Fig. 1. And conditionally speaking the nineteenth century is a century of mechanisms, twentieth is a century of machines; twenty-first is a century of systems.

3 Integration: The Theory of Mechanisms and Machines— Special Disciplines—Diploma—Scientific Work

TMM course is the first discipline of curriculum in which course design is performed. In the process of organizing of the course design is important to form a comprehensive approach for solving engineering tasks for trainees, therefore it is advisable to practice perform the complex project of the technical object (TO) from the field of future professional work of students with stage design in the TMM course, then in the MP course and then in the PD. Complex project can be development in special courses in further. Parallel study of course "Information technologies in designing

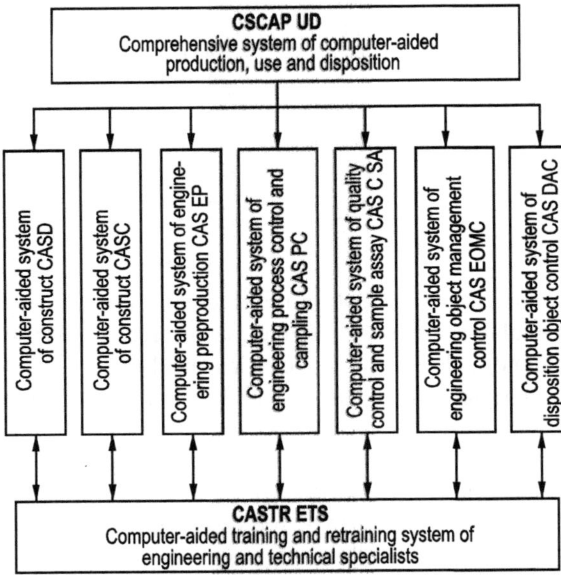

Fig. 1 The system of automation of production

of mechanical systems" in which AutoCAD, KOMPAS, VSE, DINAMIC, APM WinMachine, and other systems are studied allows performing the course design in computer-aided procedures [3].

Tasks on course design must satisfy future profession of students. For the training of mechanic engineer of maritime fleet as tasks are slider-crank mechanisms which are in the base of units of marine equipment: main ship engine, diesel generating set, piston pump, blowers and others. Kinematic scheme of slider-crank crosshead type mechanism of one cylinder of marine engine of KSZ type and its visualization in VSE system are represented on Fig. 2. The geometric analogues of kinematic parameters which depend on structure and geometry of mechanism are identified in this system too.

For the calculation of strength and design of machine elements necessary to perform an initial investigation of the dynamics for the simple dynamic model: a distributed variable mass, rotating around a fixed axis under the influence of variable torque. Research of this mechanism: structuring of mechanism scheme, kinematics research, definition of parameters of simple dynamic model on the initial stage of design, solution of the equation of motion, definition of unbalanced forces and moments of inertia, kinematic parameters of the points and links of mechanisms taking into account the masses of links and the forces, forceful calculation are performed with DINAMIC system [4].

At this stage design of technical system is performed on the stage of structural and parametric synthesis. Research results on this stage allow choosing status of links of mechanism for further stress calculation and detail and components design when the greatest forces and moments act. This calculation is performed with the

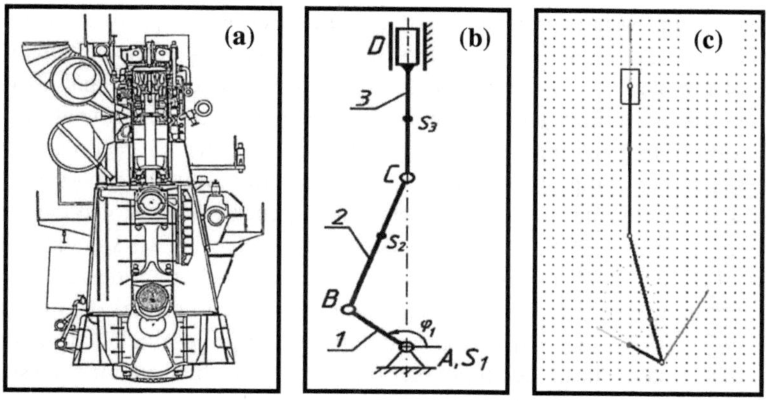

Fig. 2 Design of marine engine of KSZ type. **a** structural plan, **b** kinematic scheme, **c** scheme animation in editor VSE

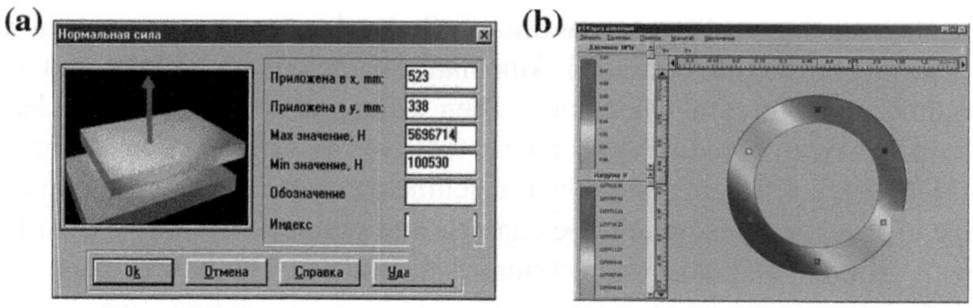

Fig. 3 Structural model and results of checking calculation of flange butt joint in APM system: **a** basic data for checking calculation, **b** pressure distribution in interference surface and load distribution on screw bolts

use of integrated APM WinMachine system. The system allows performing all the engineering designing and verifying calculations, construction of the majority of units and details mechanisms.

Some results of mechanism research in APM WinMachine system are represented on Figs. 3 and 4.

It is reasonable to continue designing of technical object for example design of the ship equipment in special disciplines and then in diploma design. The use of professional CAD in the training process should be an integral part of training of engineers

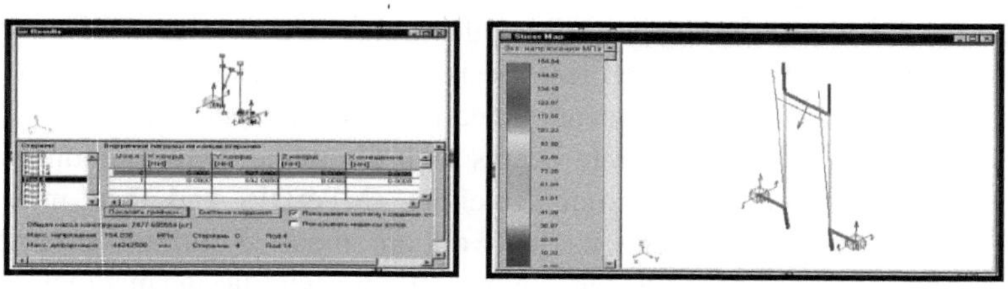

Fig. 4 The screen of calculation results and voltage and strain card in cranked portion of a shaft

and become the basis of innovative technologies and the integration processes in engineering education. Future engineers have to get an idea of the complexity and variety of the methods for tasks calculation while machine design and also about functional responsibilities and relationships between different departments and individual posts of engineering services of enterprises. In order for best realization of these tasks and teaching of future engineers to use effectively the modern software packages (MSP) during the education process of general professional engineering disciplines it's necessary to bring into use the simulation exercises simulated design process of technical object (TO), the choice of a rational variant design, technical and economic workup of various variants. It is advisable to perform not individual course projects on various courses, but a complex course project included workups on various disciplines. The model of production enterprise with all engineering services is formed within the university.

Course design of students is organized on the principle of "team training", when there are 4–5 students in each team design one type of actuating unit and choose different types of engines. First of all, students work with the structural scheme of the mechanism, and change different parameters (links size, masses, moment of inertia, power load and others), determine the kinematic characteristics of links, and perform the dynamic study, forceful calculation. Then while practicable design students can choose various types of reducing gear, gear-tooth forms, different connection of gear wheels with shaft, different materials of machine parts, etc. As a result, one team on the stage of structural and parametric analysis and synthesis designs 4–5 variants of gang with different technical and economic parameters. The best of these variants is subjected of detailed design on the stage of stress calculations and design with paperwork by means ***AUTOCAD.*** With such organization of design students work with the elements of professional packages: data banks (by parameters of standard knots—electric motors, bearing parts, coupling groups and others); with graphical databases—drawings of typical details, shafts, gear wheels, assembly components from it; with ready-made graphic calculation. They have the opportunity to see how change one of the parameters can affect on the entire structure as a whole. During university education students independently have the unique opportunity to create the element of the technical system from start to finish in full-scale. They are developmental engineers, draftsmen, process planners, product engineers and embody in course project their technical solutions like kinematic scheme or finished drawings with the methods of parts production. There is the opportunity to design not according the available model (prototype) but on the base of the physical principle of operation of the facility and to get innovative and original solutions. The use of computer-aided design systems allows setting and solving the multicriterion tasks with a numbers of control parameters in course project. Besides that this form of organization of a course design stimulates development of creative abilities of students.

Professional engineering training begins with discipline TMM, continues with courses Machine Parts and Design Principles (MP&DP) and then in special disciplines. The first skills to engineering work are imparted during course design. Realization of idea of complex design requires serious scientific and methodical studies, creation of teaching materials on modern information level. This allows training qualified engineers who master the modern tools of design automation and at the same time it's helpful for process of promotion of CAD/CAM/CAE systems in production. Today such skills are one of the criteria of competitiveness among graduates of Technical Universities on labour market.

4 Computer and Mechanical Packages

While studying TMM the experimental (laboratory) work are developed, a large amount of calculation and graphical operations during the work on course (education) projects, however experimental, graphical and analytical methods of engines research are organically matched in all modules and procedures. All of this creates a necessity for computer and mechanical complexes on OMM (CMC_TMM), based on computer technology of training [5].

CMC_TMM is a set of modern laboratory facilities, tools of training organization and training of the information environment, professional software complexes for solution of applied problems. The structure of CMC_TMM:

1. Special purpose on-line computer class which is equipped with modern computer equipment for training in computer technologies.
2. Special purpose laboratory which is equipped with laboratory benches and facilities with modern tools of running of experiment, its elaboration and realization of remote access.
3. Computer-aided training organization system (CTOS).
4. Information environment of discipline (IE_D).

The most difficult element of CMC is modern laboratory setups. Training simulators on course theory of mechanisms and machines (TMM) were created in the mid of twentieth century and unfortunately now they're worn-out and obsolete. There is a necessity to create fundamentally new installations.

In MSUN named after adm. G. I. Nevelskoy there was a first experience to realize elements of computer and mechanical complex on theory of mechanisms and machines (CMC_TMM) on the base of computer-aided training organization system "COBRA" which allow organizing class hours and independent work of trainee in computer technologies.

First laboratory work "Structural analysis of the mechanisms" is performed on the benches. Multilink mechanism (model or map case) is offered for structural analysis. First of all the structural scheme of mechanism is drawing, motion freedom of mechanism is determined; formula of the mechanism structure is written, class of mechanism is registered, the length of links is defined, coordinate system is chosen, coordinates of fixed points and position of fixed guide are determined. These data are necessary for simulation of mechanism scheme on the base of generalized structural modules in "VSE" and "DINAMIC" systems during further laboratory works and course design performing. The model of moulder-planer and its dynamic visualization in VSE system are represented on Fig. 5.

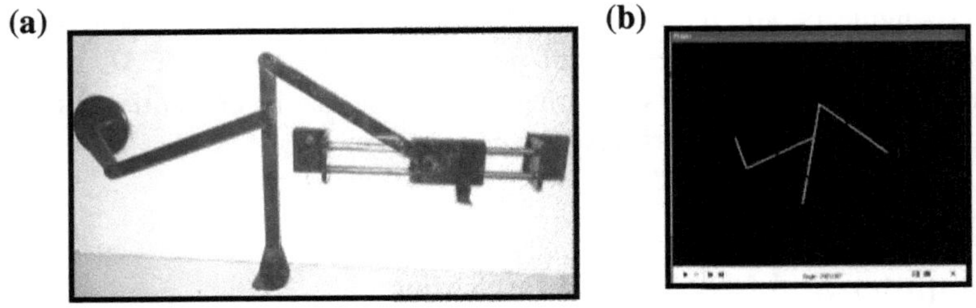

Fig. 5 Moulder-planer mechanism: **a** model, **b** dynamic visualization in VSE system

Further laboratory works in computer technologies can be performed on the example of the same mechanism but it is advisable to perform the mechanism scheme prescribed for course design. CMC is described in detail in paper [5].

CMC on the base of CTT allows substantially changing forms and methods of training, includes advanced methods of information presentation, methods of development of the creative abilities of the person, realization of idea for virtual education on engineering disciplines. It is advisable to intensify efforts on development of international integration in development of CMC and using of CTT, to create unified educational information environment and infrastructure on certain disciplines, such as TMM, or other engineering disciplines.

5 Conclusion

Integration of disciplines and relation between the various departments is often difficult because isolation of certain disciplines, various techniques of training, different levels of computer technology usage. However, the integration of training procedures on the basis of computer technology is inevitable. Single environment CAD is a part of CMC and creation of such complexes should be performed on the basis of projects that integrate scientific, technical, methodical and human resources of several teams from various universities, and possibly countries. The final result of such a project should be the CMC in form of laboratories with a set of laboratory and computer equipment and methodical (computer) software, and also computer disk and hard copy of instruction for use.

The creation of CMC requires significant financial embedding and integration of intellectual resource of a region, country and possibly of several states. This project is a long-term one and can be realized step by step. Nowadays there are several international projects of two-side agreements for cooperation are offered. Within the framework of these trends Engineering Computer Centre Vector offers to realize a joint project—Development of computer and mechanical complex on course theory of mechanisms and machines (CMC_TMM). After realization of this project and basing on it, it will be possible to realize other projects: CMC on Mechanics (CMC_M), applied mechanics (CMC_AM), machine parts and design basis (CMC_ MP T DB) and on other engineering disciplines. The education facilities and their profound part (filling) can be submitted in languages of partners to of this project. The use and development of CAD tools (with the participation of students) to create an information support of training process and production (databases, libraries, typical design and technological solutions and their elements). Development of information tools of training is a new way of transfer of knowledge which is particularly important in a well-known trend of recent years such as specialists ageing of high school.

References

1. Kuzlyakina, V.V.: Information environment for mechanism and machine science course. In: Proceedings of the 11th World Congress in Mechanisms and Machine Science, vol. 2, pp. 685–688. China (2004)
2. Kuzlyakina, V.V., Brazhnik, L.: Use of professional cad systems in educational process. In: Materials of Symposium IGIP 37 Engineering Competencies—Traditions and Innovations, pp. 103–111. Moscow (2008)
3. Kuzlyakina, V.V., Kholina, N.N.: Computer-aided course project work in mechanism and machine science course. In: Proceedings of the 11th World Congress in Mechanisms and Machine Science, vol. 2, pp. 669–672. China (2004)
4. Kuzlyakina, V.V., Slepenko, U.N.: Forceful calculation of linkage on the basis of generalized structural modules. In: The First IFToMM Asian Conference on Mechanism and Machine Science, p. 6. Taipei (2010)
5. Kuzlyakina, V.V.: Computer and mechanical packages on engineering disciplines. In: Materials of Design Baseline and Machinery—XXI Conference, Orel (2007)

Part II
Mechanism and Machine Science in the Engineer Program: Methodology

Learning Through Competitive Environments: Use of Robotic Simulation Environments Combined with Economic Considerations

J. L. Suñer Martínez and F. J. Rubio

Abstract This paper presents a virtual robotic manufacturing environment created with the program about Robotic Simulation Environments called Grasp10©, which is used in teaching Robotics subjects. In this virtual environment, students must conduct a competition consisting of improving the working and financial performance of the line by introducing the changes they consider they will have positive impacts. The working of the line is based on motion and tasks planning of the different elements that take part in it. Economic performance is based on the application of different quantifiers as investment costs and benefits of exploitation. The highest grade when evaluating the solution will be for students who provide the most effective solutions from both the working and economical point of view.

Keywords Robotics · Simulation · Competition · Investments · Benefits

1 Introduction

Robotics in Manufacturing is a course that aims to introduce students to the knowledge of industrial robots, its use and implementation. The current curriculum consists of theory (by conventional means) and laboratory work.

Due to the nature of the industrial robots, to do practices with real elements is unfeasible (complex facilities, expensive and heavy security). The use of simulation software serves largely to overcome these difficulties, allowing the student to work with the same elements, concepts, and restrictions on the actual facilities, but at a much more affordable cost and without risk to their safety.

J. L. Suñer Martínez (✉) · F. J. Rubio
CITV, Universitat Politècnica de València, Valencia, Spain
e-mail: josuner@mcm.upv.es

F. J. Rubio
e-mail: frubio@mcm.upv.es

J. C. García-Prada and C. Castejón (eds.), *New Trends in Educational Activity in the Field of Mechanism and Machine Theory*, Mechanisms and Machine Science 19,
DOI: 10.1007/978-3-319-01836-2_7,

The course aims not only to introduce students to the knowledge of industrial robots and its applications but also to know the problem of the introduction of robots in industry (see [8, 9]). Simulation is a powerful tool in this field, since you can work in three-dimensional virtual environments with the option to control the motion of robots and other elements of a manufacturing line, such as automated guided vehicles, machine tools, conveyors, etc. and it enables to try different arrangements, lay-outs, programming strategies and resource management tasks so that simulation makes other methods of planning fall far behind in performance (see [3]). In the teaching of this course, the students will work with the kinematic structure of the robot and will learn the interactive handling of the robot, the setting of trajectories both point to point and continuous rectilinear. Short robot motion programs are made in gesture mode, and an element is incorporated to the robot end-effector so that the variation of the kinematic chain with this incorporation is checked (see [4–6]).

Economic factors are also taken into account as key aspects of the competition so that the basics of the motion planning, lay-out and economic issues will be combined.

In this paper, students must conduct a competition consisting of improving not only the working but also the economic performance of a robotic line by introducing the changes they consider they will have positive impacts. Therefore, the students will work on three fundamentals pillars when planning the general working of a robotic line: one is the motion planning, the second is the task planning (see [10]) and finally the economic performance. The economic performance is based on the application of different quantifiers as investment costs and benefits of exploitation. The highest grade when evaluating the solution will be for students who provide the most effective solutions from both the working and economical point of view.

2 Competition

Work will focus on how to improve the economic performance of a robotic manufacturing line without forgetting the motion and task planning. This line is equal for all participants.

Students will focus not only on programming the robot motion and task synchronization but also on economic issues related to productivity.

More specifically, they will look for the number of manufactured workpieces per hour in order to amortize the initial investment and the time needed at a constant rate of working to amortize that initial investment. The higher these indices are, the higher the grade of the students will be. In short, productivity, as a numerical value and therefore target is the strongest arguments to set the mark, although the quality of programming or innovative elements and ideas will also be consider in order to set the student's the final mark.

Economic factors will have a significant importance: it could determine main traits such as the profitability of the line, the benefit obtained, the return time of the investment, etc.

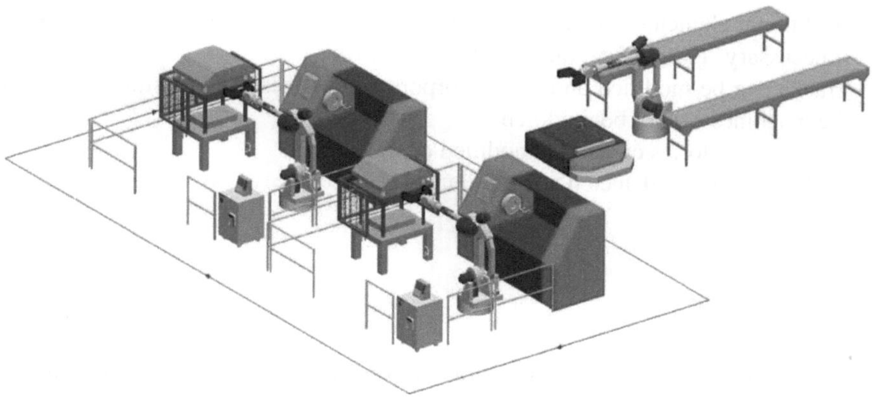

Fig. 1 Robotic manufacturing line to start the competition (baseline)

One of the keys when considering this competition is that the problem has not a unique solution, but it can have many and all can be correct in varying degrees. In this way, each student can find and develop his own solution up to the end.

3 Playground

Figure 1 shows the basic robotic line to be improved.

The initial conditions to do the work are as follows. The line consists of two blocks:

1. An input-output station composed of a robot and two conveyors.
2. Two working stations composed of a press, a lathe and one robot in charge of the maintenance of both machines.

There is also one self-guided vehicle (AGV) to haul the parts from the input-output station to the working stations.

The manufactured workpieces will suffer operations in the press and the lathe respectively (see [2, 7]). The tasks have been highly simplified.

4 Rules of the Game

These rules are:

1. Location of conveyors cannot be modified.
2. Location of workstations 1 and 2 cannot be modified, nor the side of the access to them. The location of the access can be varied but not the side.
3. The network of the auto-guided vehicle cannot be changed in the basic shape.

4. Additional elements can be incorporated (other machines, vehicles, robots, etc.) if necessary. The original ones cannot be removed.
5. Tracks can be modified and also incorporate new ones. The existing tracks for the machines cannot be modified.
6. Robots trajectories cannot be modified or substituted.
7. Collision must be detected and eliminated.

5 The Referee

Teacher is the referee of the competition. He verifies that all the students meet the rules and he is consulted to determine if their strategy is correct or feasible and also advise them.

When marking the works, the main criterion used is the greatest productivity of the robotic line. He also take into account others factors such as quality of programming, the ability to avoid obstacles, the suitability of the layout and the best use of the workspace of the robots, which can modify substantially the grade obtained finally, either upward and downward.

6 Economic Planning

The ease of programming GRASP10 enables to introduce the economic analysis as a key aspect that students must consider when undertaking their work.

Specifically, students are given a procedure which has implemented the method of economic analysis to calculate the payback period, which is the period of time from the beginning of the project until the net cash flow or the accumulated cash is greater than zero. This period should be as short as possible, because if the payback period is near or above the lifetime of the project, the project will not be worth or the initial investment will not be recovered.

The payback period is usually measured in years and it is given by the value of n that satisfies this equation (see [1]):

$$\sum_{j=0}^{n}(R_j - C_j) = 0 \tag{1}$$

where R_j and C_j are the income and estimated costs for year, respectively. The investment cost should be included and it is the value C_0 when $R_0 = 0$. It is simplified by considering the steady flow over the life of the project and n can be deduced from the equation:

$$-CI + n \cdot (R - C) = 0 \tag{2}$$

Since the initial expenses are CI and $(R - C)$ the annual cash flow, solving for n:

$$n = \frac{CI}{(R - C)} \tag{3}$$

Here CI is the unitary price of the machines, robots and AGVs multiplied by their respective units. C is the estimated annual cost and it is given by the consumed workpieces (the same number that the manufactured workpieces) multiplied by the price per unit and some fixed costs due to labor, which depend on the facilities you have.

R is the estimated annual income and it is given by the number of manufactured workpieces and manufactured in multiplied by the unitary sale price.

The procedure can calculate payback period considering that there can be 3 programs of 1, 2 and 3 shifts per day, i.e. 5, 10 and 19 weekly shifts.

Besides, the procedure has been completed with the estimated selling price that must have the workpiece to have a payback period of 3, 5 and 10 years, assuming the same work program. The formula to determine that is:

$$PPz = \frac{1}{PzAn} \cdot \left(C + \frac{CI}{n}\right) \tag{4}$$

where PPz is the selling price of the workpiece and $PzAn$ the number of workpieces produced annually.

With all this, the tasks to be done by the students are to improve the behavior of robots, to plan the motions, to improve the performance of the facility, job scheduling, and the obtaining of the best economic performance by means of a convenient economic planning.

7 Results of the Competition

This section shows two examples (minimum framework and maximum framework) created by students, in which they can compare their results against the baseline.

The following prices and costs are considered as common start (Table 1):

Items in the 3 manufacturing lines can be seen in Table 2.

Table 1 Items and prices

Item	Robot	Lathe	Press	AGV	Worker	Raw workpiece	Manuf. workpiece
Price (€)	60.000	18.000	6.000	60.000	50	50	55

Table 2 Number of units in three different environments

	Baseline	Min. framework	Max. framework
Robots	3	3	8
Lathes	2	2	6
Presses	2	2	6
AGVs	1	2	10
Workers	1	1	3
Initial costs	288.000€	348.000€	1.164.000€

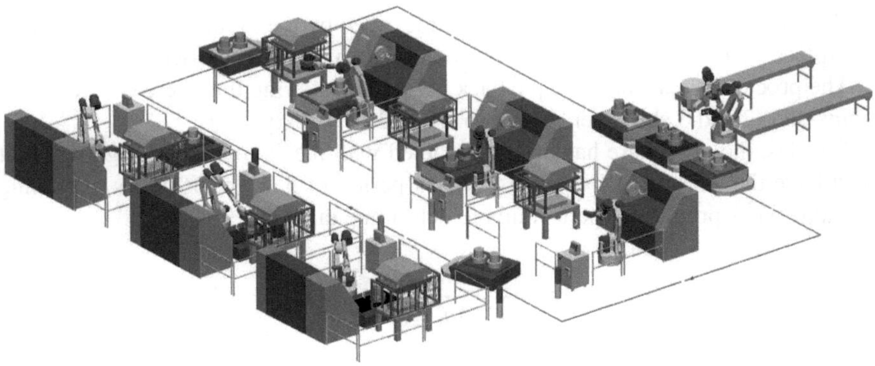

Fig. 2 Maximum Framework

In Fig. 1 (above shown) can be seen the robotic baseline and in Fig. 2 can be seen the maximum number of item that there can be in the robotic manufacturing line. There may be up to 8 robots, 6 lathes, 6 presses and so on.

Results: In Table 3, it can be seen the payback period in years depending on the characteristics of the robotic manufacturing line and the number of shifts per year.

It can be seen as the number of years to payback the facility decreases as the number of shifts increases.

Furthermore, the number of years to payback the facilities also decreases if additional elements are added to the manufacturing line (represented by Minimun and Maximum framework).

Table 3 Number of periods (in years) to payback workpieces manufactured

	Baseline	Min. framework	Max. framework
Workpieces per hour	20,534	54,282	178,760
Payback period with 1 shift (years)	1,436	0,6301	0,6395
Payback period with 2 shifts (years)	0,718	0,315	0,320
Payback period with 3 shifts (years)	0,378	0,166	0,168

Table 4 Number of workpieces manufactured to payback

	Baseline			Maximum framework		
	3 years	5 years	10 years	3 years	5 years	10 years
1 shift/year	52,56€	51,43€	50,90€	51,42€	50,63€	50,38€
2 shifts/year	51,66€	50,98€	50,66€	50,73€	50,42€	50,27€
3 shifts/year	50,98€	50,65€	50,49€	50,42€	50,27€	50,19€

The most favorable situation in terms of payback time is given with Minimun framework with three shifts per year. In this case, the initial investment of 348,000€ is amortized over 0.166 years, i.e., 1 month and 30 days

In Table 4, the price of a workpiece can be seen, depending on the working shifts per year and the payback period in years for the robotic baseline and the Maximum framework.

It can be seen as the price of the workpiece decreases when the years considered that the production line will be operational is increased (3, 5 or 10 years). This trend is true both for the baseline and for the Maximum framework.

On the other hand, it can be seen as the price of the workpiece decreases as the number of shifts per year increases. This trend is true both for the baseline and for the Maximum framework.

Furthermore, it can be seen as the price of the workpiece also decreases if additional items are added to the production line, represented by the baseline and Maximum framework for equal periods of years.

The cheapest item is produced for Maximum framework, considering three shifts per year and a payback period of 10 years. In this case, the initial investment of €1,164,000 is amortized over 10 years, working in three shifts per year with a price of €50.19 piece.

8 Conclusions

These practices show that competition can be applied in teaching. In trying to obtain the best marks, the students make a deep analysis of the working of the manufacturing line and they obtain very good solutions that improve the behaviour of the baseline.

In this paper, the role that economic performance plays in the design of a robotic manufacturing line has been analysed. It has received a central attention. Simultaneously, the usual motion and task planning have been also considered as indispensable part of the same analysis.

The benefits that can stand out of this experience are:

1. Competition strengthens creativity in seeking solutions to the problem. All years that this resource has been used, there have been students who have created new strategies to improve the initial situation.

2. Students compare their strategies with each other and it may appear challenges between them, so they are longing to improve their solutions.
3. Programming is a subject that is not always appreciated by the students. However, when confronted with its obligatory and also to improve it, at the end of these practices the reluctance to use the programming has been overcome, which can be transferred to other disciplines that also uses programming.

Some inconveniences that have arisen over the years in which this experience has been developed are few and are concentrated in the following two:

1. Some students may spend more time than advisable in an effort to get the highest score possible.
2. If students have a poor programming background, the time resolution of the problem can be increased considerably.

As for the economic study is concerned, the result is that cheapest workpiece will be produced when the number of items (robots, machines, etc.) is maximum, considering three shifts per year and a payback period of 10 years. In the studied case, the initial investment of €1,164,000 is amortized over 10 years, working in three shifts per year with a price of €50.19 piece.

As a final conclusion, we can say that the use of a competition for teaching has been very positive in "Robotics in Manufacturing". In the near future, one could use the program Grasp10© to introduce random factors (defective, damaged machinery) that would serve to test the robustness of the management and the ability to response against interference of the modified installation created by students.

Undoubtedly, with time enough, the competition could expand the number of workpieces manufactured and be much more rewarding for those students who performed it.

References

1. Barrientos, A., Peñín, L.F., Balaguer, C., Aracil, R.: Fundamentos de Robótica 2^a Ed. Ed. McGraw-Hill/Interamericana de España, Madrid, España (2007)
2. Boothroyd, G.: Assembly Automation and Product Design, Second Edition. CRC Press, Taylor & Francis Group, Boca Raton, Florida (2005)
3. BYG Systems Ltd.: GRASP10 User Manual First Edition. Nottingham (2006)
4. Craig, J.J.: Introduction to Robotics, Ed. Addison-Wesley Publishing Company, U.S.A. (1989)
5. Fu, K.S., González, R.C., Lee, C.S.G.: Robótica: Control, Detección, Visión e Inteligencia. Ed. McGraw Hill, Madrid, España (1988)
6. Megahed, S.M.: Principles of Robot Modeling and Simulation. Ed. John Wiley and Sons Ltd., Chichester (1993)
7. Ray Asfahl, C.: Robots and Manufacturing Automation, Ed. John Wiley and Sons Ltd., New York (1993)
8. Suñer, J.L., Mata, V.: Utilización de Programas de Simulación en la Docencia de la Ingeniería Mecánica. Actas del X Congreso de Innovación Educativa en las Enseñanzas Técnicas, Valencia (2002)

9. Suñer, J.L.: Apuntes prácticas de Robótica en Producción. Servicio de reprografía UPV, Valencia (2008)
10. Suñer, J.L., Rubio, F.J., Mata, V.: La Competición como Herramienta Docente. Un caso con Simulación de Entornos Robotizados. XIX Congreso Universitario de Innovación Educativa en las Enseñanzas Técnicas - XIX CUIEET, ISSN 978–84-694-2160-4, (2011)

Service-Learning Projects Based on Dynamic Documentation in Engineering Colleges

J. M. García-Alonso, E. Soriano, I. García-Vicario and H. Rubio

Abstract In recent years, engineering colleges have started to integrate service-learning projects into many of their required courses. In the literature, there are many important definitions for service-learning pedagogy, including reciprocity, reflection, coaching and community projects. This paper presents a service-learning pedagogy implementation program for educating engineering undergraduates to solve real-world problems. Three project models for mechanical engineering, ranging from a single semester to a full calendar year, form the basis for the analyses presented. Finding appropriate community partners and projects are critical for a successful experience for all involved parties. From the results of this analysis, it is possible to conclude that, through these projects, students acquired and applied the competencies in the defined curriculum.

Keywords Service-learning · Dynamic documentation · Community engagement · Mechanical design

---

J. M. García-Alonso(✉)
Salesianos Atocha, Madrid, Spain
e-mail: jesusmanuel.garcia@salesianosatocha.es

E. Soriano · I. García-Vicario and H. Rubio
University Carlos III of Madrid, Madrid, Spain
e-mail: esoriano@ing.uc3m.es

I. García-Vicario
e-mail: igvicari@ing.uc3m.es

H. Rubio
e-mail: hrubio@ing.uc3m.es

J. C. García-Prada and C. Castejón (eds.), *New Trends in Educational Activity in the Field of Mechanism and Machine Theory*, Mechanisms and Machine Science 19,
DOI: 10.1007/978-3-319-01836-2_8,

1 Introduction

In recent years, engineering colleges have integrated service learning into many of their required courses.

There are several different definitions for service learning in the existing literature. Jacoby [1] defines service learning as the integration of academics with community service in credit courses, with the key elements such as reciprocity, the voice of reflection, coaching and community projects. To Duffy et al. [2], service learning is a 'practical' approach to learning in which students achieve academic goals in a credited course for meeting the real needs of the community. Service learning is defined by Bringle et al. [3] as a credit-based, course-supported educational experience in which students participate in an organised service activity that meets defined educational outcomes and addresses the needs of the community, while requiring students to reflect on the service activity to gain a greater understanding of course content. Several service projects in engineering courses does not meet the full definition of service-learning projects [4].

Several research projects have proposed training programs based on service-learning pedagogy. Bielefeldt et al. [4] proposed five model projects that have been tested in universities in the United States. Duffy et al. [2] proposed 35 different courses based on service learning that can be implemented in engineering courses. Hulman et al. [5] discussed the design of a bioengineering course within a project based on service-learning pedagogy, demonstrating that a combination of challenging engineering design projects and long-term service to the community proves to be extremely successful. Coyle et al. [6] employed service-learning programs in which students earned academic credit for their participation in working groups. These working groups resolved technical problems proposed by not-for-profit organizations. Tsang et al. [7] implemented a service-learning program in an initial course of mechanical engineering. During the 3 years of their study, service-learning pedagogy proved to be an important tool in teaching and practice of engineering that enhanced teamwork and human relations. Recently, Oakes et al. [8] presented a methodology that included training models based on service-learning pedagogy. In these models, instructors must devote both time and attention for developing relationships with partners prior to the beginning of the course and when following up to help ensure optimal outcomes for the partners.

This paper presents the service learning pedagogy implementation in educating engineering undergraduates, at Colegio Salesianos Atocha (CSA) and University Carlos III of Madrid (UC3M), to reach the actual societal challenges. Three different projects form the basis for the analyses presented. These projects are summarized in Table 1.

Finding appropriate community partners and applicable projects are critical for a successful experience for all involved parties. Students form teams of three or four and select the project on which they want to work.

From the results, it is possible to conclude that students acquired and applied the competencies in the defined curriculum; but it was also discovered that the projects

Table 1 Summary of project models incorporating service learning

Project name	Roller bearing bench device	Automatic collet-chuck	Roller compactor	Central catheter of peripheral access	Central catheter
Department	Mech.	Mech.	Mech.	Mech.	Mech.
Location	UC3M	UC3M	CSA	CSA	CSA
Majors of participating students	Bachelor mechanical engineering	Bachelor mechanical engineering	High vocational technician	High vocational technician	High vocational technician
Number of students enrolled per project	4	4	5	5	5
Number of faculty	2	2	4	4	4
SL clients/community partners, tutors	2	2	2	2	2
Special support equipment	Yes	Yes	No	No	No
Frequency of client interactions	Once a month face to face meetings	Once a month face to face meetings	Once a month face to face meetings	Once a month face to face meetings	Once a month face to face meetings
Lifecycle	Yes	Yes	Yes	Yes	Yes
Sustainability	Yes	Yes	Yes	Yes	Yes
Eco-design	Yes	Yes	Yes	Yes	Yes
S.L. client	Small companies	Small companies	Small companies	Non-profit	Non-profit

resulted in an additional workload for the college staff. The easiest way to improve the efficiency in service-learning pedagogy is by providing training to the university offices that will facilitate the interactions with service-learning clients.

2 Teaching-Learning Methodology

2.1 Finding Project Partners and Sequence of Students' Activities

Finding appropriate community partners and projects are the cornerstones for successful service-learning experiences. This process requires the identification of companies, non-profit organizations, municipalities and others interested in offering projects that can be reliably performed by freshmen and finished on time. The community partner should be available to meet the instructor, provide information and ideas prior to starting the course and be available for questions during the project; but this does not necessarily mean that they have to invest a lot of their time. However, at the time of defining students' projects, it is helpful if partner representatives can meet the students directly in order to clarify doubts.

At CSA and UC3M, projects have been identified through contacts with two small companies and a non-profit organization with limited financial resources. One of the partner companies repairs consumer machines and was interested in developing an online course for disabled workers and creating an online brochure. The other small company was interested in developing a collet chuck for turning machines. The non-profit organization was interested in developing a low-cost wheelchair and had engineers on staff, allowing students to continue the projects after their coursework was complete.

Professors, instructors and clients began by identifying the tasks to be performed by students and defined the timelines in which the tasks would be completed. Later, students developed the technical proposals which defined their scopes of the work and included their statement of qualifications among other things. After that, students made poster presentations to the partner companies, conducted feasibility studies and provided alternative assessments and/or created a prototype (see Fig. 1).

2.2 Forming Teams for the Service-Learning Projects

Teams comprising three to four students appear optimal. Students' preferences were the primary criteria used to form teams. Service-learning projects were among the top three choices of the students.

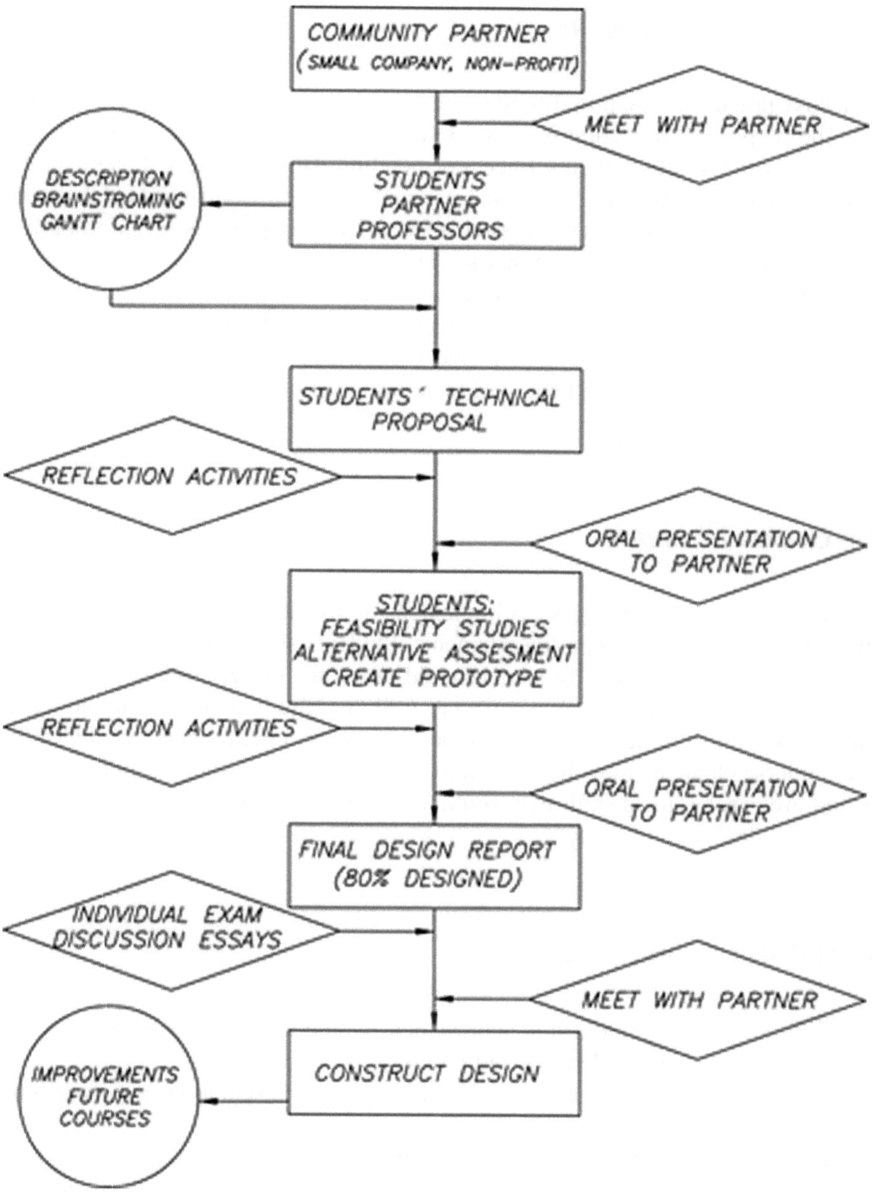

Fig. 1 Sequence of Students' activities

2.3 Students' Reflections

Reflective practice is a required component of rigorous service-learning experiences [4]. The purpose of reflection is to promote learning about the larger social issues behind the needs to which their service is responding. This learning includes a deeper understanding of historical, sociological, economic and political contexts of the needs or issues being addressed [1]. Proposed service-learning projects include multiple student reflection elements, as summarised in Table 2. The quantitative and qualitative components of the data were analysed separately and the data derived from the focus groups were classified as strong, moderate or weak. Students' assessments of a given criterion were classified as strong if members of at least four of the seven focus groups believed there was a positive benefit of their participation in service learning in this course [5].

2.4 Course Elements

When service-learning pedagogy is implemented, instructors and professors should take into account a few important elements. Students should develop their design in one semester or in 1 year. Students should finish any project they started so that the project can be a rewarding experience. Additionally, students should pay close attention to the sustainability of the project and to its environment and lifecycle. These course elements must be taken into account in order to successfully achieve the required professional skills that would have to be implemented at the conclusion of their studies.

3 Student Learning Outcomes

In order to assess each student's achievement of the learning goals, we use two associated percentages (individual and group).

The group percentage was assigned during the oral presentations once a month, as shown in Table 3.

The individual component consisted of two assessments, one for competences achieved by means of the project and another for skills and abilities to solve or manage similar projects. We performed an initial assessment at CSA. In this assessment, a student disassembled the machine, and was then encouraged to reassemble it. The maximum allowed time was 10 min. At the end of the course, the same assessment was carried out. The results of both assessments (initial and final) are summarised in Table 4.

Table 2 Summary of students' reflections in meeting ABET objectives

Criterion	Quantitative CSA	Quantitative UC3M	Qualitative CSA	Qualitative UC3M
An ability to apply knowledge of mathematics, science and engineering	Moderate	Strong	Moderate	Moderate
An ability to design and conduct experiments, as well as to analyse and interpret data	Strong	Strong	Strong	Strong
An ability to design a system, component, or process to meet desired needs	Strong	Strong	Strong	Strong
An ability to function in multidisciplinary teams	Moderate	Strong	Strong	Strong
An ability to identify, formulate and solve engineering problems	Strong	Strong	Strong	Strong
An understanding of professional and ethical responsibility	Strong	Strong	Strong	Strong
An ability to communicate effectively	Strong	Strong	Strong	Strong
The broad education necessary to understand the impact of engineering solutions in a global and societal context	Strong	Strong	Strong	Strong
A recognition of the need for an ability to engage in lifelong learning	Strong	Strong	Strong	Strong
A knowledge of contemporary issues	Moderate	Strong	Strong	Strong
An ability to use the techniques, skills and modern engineering tools necessary for engineering practice	Strong	Strong	Strong	Strong
An ability to make measurements on and interpret data from systems	Strong	Strong	Strong	Strong

Table 3 Group assessments

Project	Technical proposal (%)	Solution (%)	Design (%)	Prototypes (%)	Oral presentation (%)	Partner's assessment (%)
Bearing bench device	5	25	20	30	10	10
Automatic collet-chuck	5	25	20	30	10	10
Roller compactor	5	25	20	30	10	10
Catheter of peripheral access	5	25	20	30	10	10
Central catheter	5	25	20	30	10	10

Table 4 Group assessments

Project	Location	Student	Reassembled machine time (initial)	Reassembled machine time (final)
Bearing bench device	UC3M	1	12′ 32″ *	3′ 15″
		2	11′ 43″ *	3′ 00″
		3	12′ 52″ *	4′ 25″
		4	10′ 46″ *	5′ 41″
Automatic collet-chuck	UC3M	5	abandoned	5′ 35″
		6	07′ 43″ *	5′ 57″
		7	10′ 36″ *	4′ 45″
		8	12′ 09″*	6′ 32″
Roller compactor	CSA	9	11′ 41″ *	3′ 43″
		10	09′ 10″ *	4′ 20″
		11	13′ 00″*	4′ 39″
		12	12′ 17″ *	4′ 05″
Catheter of peripheral access	CSA	13	10′ 24″ *	5′ 26″
		14	11′ 02″ *	5′ 57″
		15	12′ 47″ *	5′ 48″
		16	09′ 36″ *	7′ 06″
Central catheter	CSA	17	12′ 15″ *	5′ 21″
		18	12′ 20″ *	abandoned
		19	abandoned	5′ 01″
		20	12′ 41″ *	6′ 14″

As shown in Table 4, in the early phase, the majority of the students were not able to reassemble the machines within 10 min. At the end of the course, however, students were able to reassemble the machine with shorter task times.

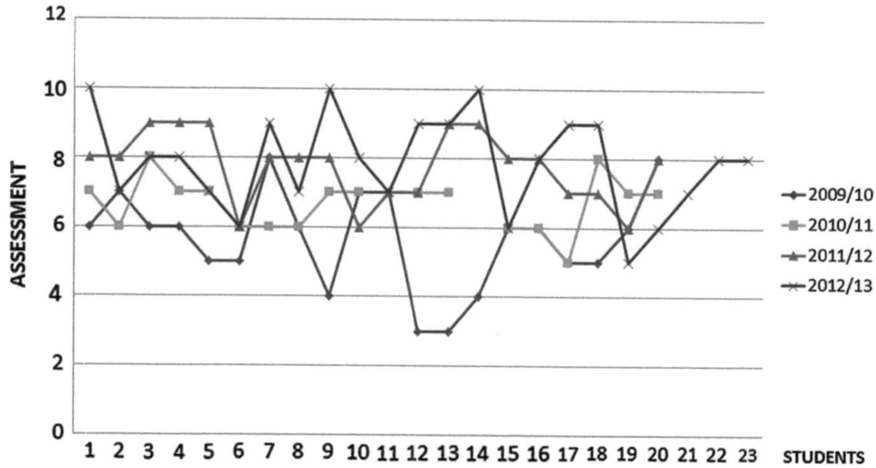

Fig. 2 Machine assembly-disassembly assessments

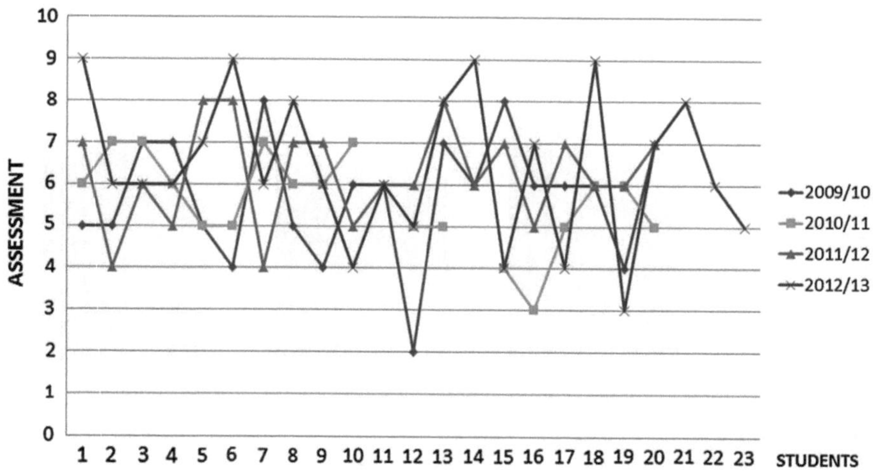

Fig. 3 Graphic representation assessments

4 Benefits of the Service-Learning Projects

As shown in Figs. 2, 3, 4 and 5, during the 2009–2010 academic year, as students worked on the same machine, there were improvements in both skills and training. It was felt, however, that some students applied themselves only minimally to the effort and simply copied the work of their peers. During the 2010–2011 academic year, when professors and instructors distributed different machines to the students, they noticed a substantial improvement in students' outcomes. Deeper analyses of the teaching/learning process, however, showed that students learned a lot about the

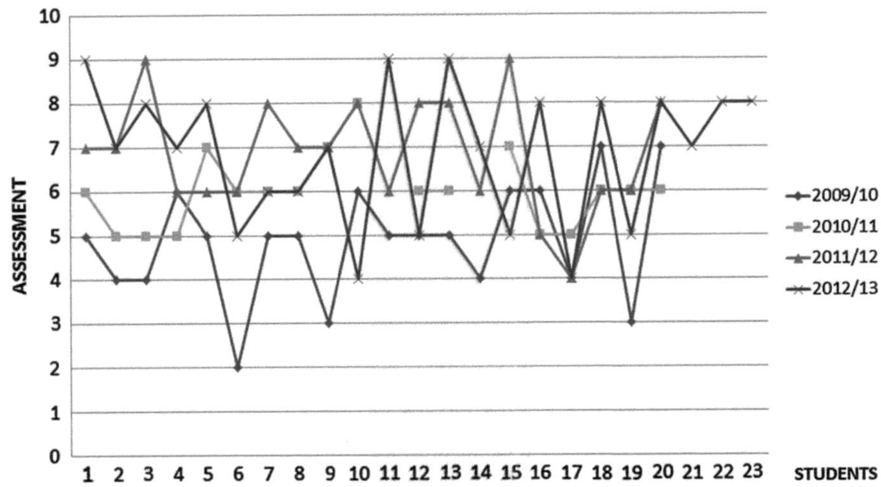

Fig. 4 Machine elements assessments

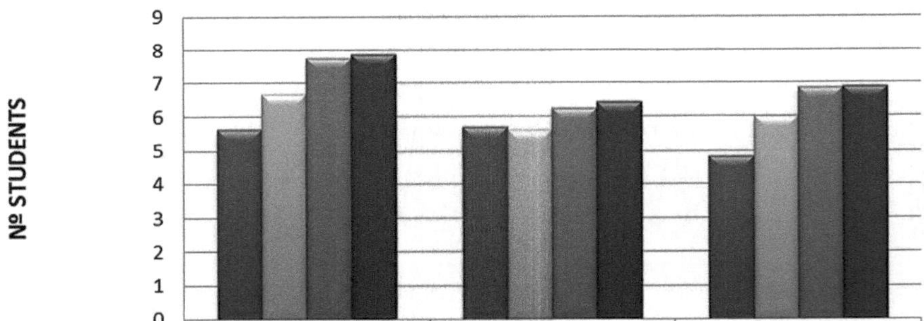

	ASSEMBLY AND DISASSEMBLY MACHINE	GRAPHIC REPRESENTATION	MACHINE ELEMENTS
Course 2009/10	5,65	5,7	4,85
Course 2010/11	6,67	5,61	6
Course 2011/12	7,75	6,25	6,85
Course 2012/13	7,87	6,43	6,87

Fig. 5 Assessment evolution

machines themselves, but they were not able to use the machines to sort, classify and manage products.

During the 2011–2012 and 2012–2013 academic years, service-learning pedagogy was implemented and included the following features:

- Projects or machines from the same community partner, so that students could work on similar projects.

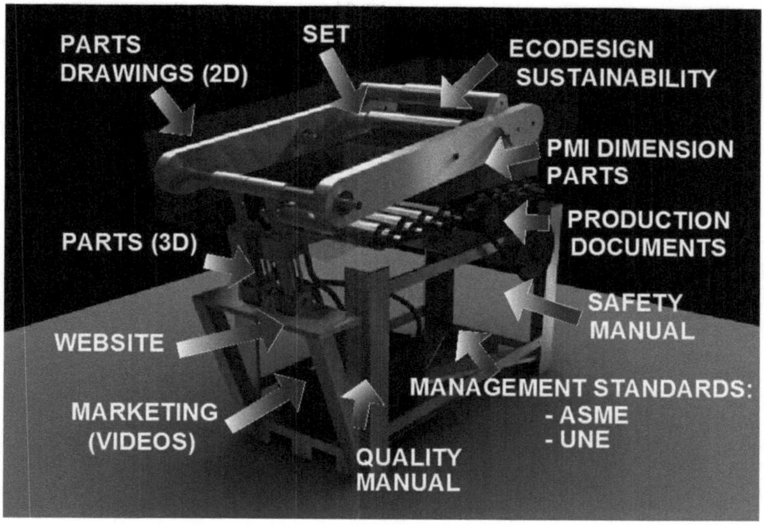

Fig. 6 Subject theory by means of service-learning projects

- Projects provided students with items that needed improvement at the company or in society.
- Learning should not be limited only to a single project but to several similar projects.

The service-learning pedagogy implemented during the 2011–2012 and 2012–2013 calendar years supplied the following improvements with respect to previous academic years:

- Higher involvement of students.
- More student interaction with industry and real-world projects (better professional integration).
- Emphasis was placed on practical work. Professors were therefore able to introduce subject theory with the application of service-learning projects (see Figs. 6, 7).
- Achievement of adequate improvements in development and evaluation of competences resulting in, improved students' subject assessments (see Figs. 2–5).

On the other hand, service-learning pedagogy requires an additional workload for the college staff, especially for professors and instructors. The easiest way to improve efficiency in service-learning pedagogy is by means of learning offices that are able to facilitate interactions with service-learning clients.

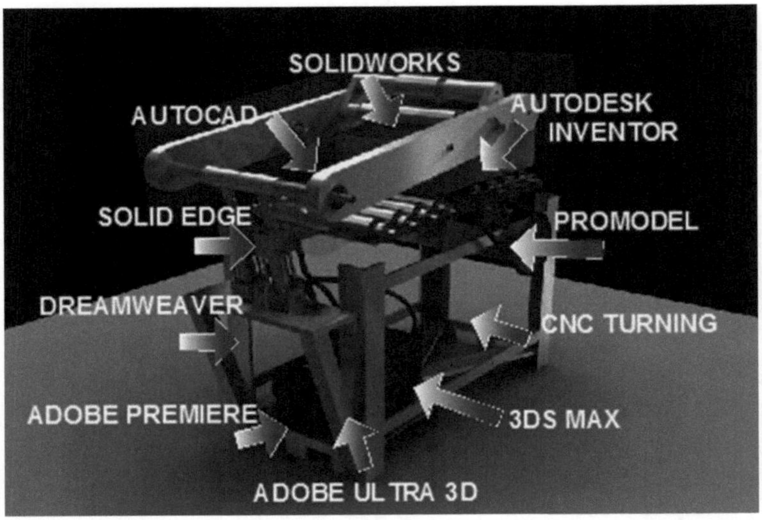

Fig. 7 Subject software by means of service-learning projects

Fig. 8 QR code

5 Conclusions

During the 2011–2012 and 2012–2013 calendar years, we implemented service-learning pedagogy in two engineering colleges: CSA and UC3M.

The pedagogical effort was not to focus on the results obtained from a single test exam but from a doing/learning methodology. Additionally, an active role was transferred to the students who were allowed to put into practice the professional skills that would have to be implemented at the conclusion of their studies. Students' teamwork skills improved and the students showed improved academic results.

Professors and instructors could implement and demonstrate the validity of their theoretical models while considering ergonomic and environmental factors, assembly resources, workplace design and safety issues. They were able to create worker training procedures and guides and to demonstrate the possibility of a new product assembly that offers both time and cost reductions.

In order to achieve an adequate development and evaluation of competencies, this proposal can be applied and studied in a program that includes subjects divided among various educational levels or that considers different curricula objectives.

Additional studies can include ways to improve the efficiency of service-learning pedagogy and reduce the workload for professors and instructors by implementing learning offices.

Figure 8 shows the QR code to see the dynamic documentation of the "roller compactor" service-learning project.

References

1. Jacoby, B.: Service-learning in today's higher education. In: Jacoby B. and Associates (eds.) Service-Learning in Higher Education: Concepts and Practices, pp. 5–13. Jossey-Bass, New York (1996)
2. Duffy, J., Barrington, L., Moeller, W., Barry, C.: Service-learning projects in core undergraduate engineering courses. Int. J. Serv. Learn. Eng. **3**(2), 18–41 (2008)
3. Bringle, R., Hatcher, J.: A service-learning curriculum for faculty. Mich. J. Community Serv. Learn. **2**, 112–122 (1995)
4. Bielefeldt, A., Dewoolkar, M., Caves, K.: Diverse models for incorporating service projects into engineering capstone design courses. Int. J. Eng. Educ. **27**(6), 1206–1220 (2011)
5. Ropers Hulman, B., Carwile, L., Lima, M.: Service-learning in engineering: a valuable pedagogy for meeting learning objectives. Eur. Int. J. Eng. Educ. **30**(2), 155–165 (2005)
6. Coyle, E.J., Jamieson, L.H., Oakes, W.: EPICS: engineering projects in community service. Int. J. Eng. Educ. **21**(1), 139–150 (2005)
7. Tsang, E., Haneghan, J., Jhonson, B., Associates: A report on service-learning and engineering design: service-learning's effect on students learning engineering design in introduction to mechanical engineering. Int. J. Eng. Educ. **17**(1), 30–39 (2001)
8. Oakes, W.P.E., Zoltowski, C.: Workshop: service-learning in engineering, technology and computing. In: 41st ASEE/IEEE Frontiers in Education Conference W3D–1, October 2011

Modern Means and Technologies of Training on Course Theory of Mechanisms and Machines

V. V. Kuzlyakina

Abstract Modern engineering education is intended to increase the natural sciences base of training, to introduce to the educational process mainly analytical methods of research and design of objects, to use the advanced and professional system for computer-aided design, and the availability of modern laboratory facilities and computer equipment. Course "Theory of Mechanisms and Machines" is a fundamental discipline in the training structure of mechanic engineer. A large amount of information, a high level of difficulty of content and different forms of training sessions forced to make a training process on modular approach. Modular approach of training process is defined not only by the complexity and volume of material, but by modern trend to the wide use of computer technologies. Computer technologies of training can change the forms and methods of training fundamentally, including the modern way of information's present, methods of amplification of person's creativity. Nowadays the task of creating of information environment on course "Theory of Mechanisms and Machines" on base of computer-aided training system is especially important.

Keywords Information environment · Creative training system · Testing · Automation of design

1 Introduction

Course "Theory of Mechanisms and Machines" is a fundamental discipline in the training structure of mechanic engineer. The model of a modern training is to develop an innovative system of training in all procedures of training on each discipline of the educational program. It lets to pass on from the reproductive cognitive activity

V. V. Kuzlyakina (✉)
Maritime State University named after adm. G. I. Nevelskoy, Vladivostok, Russia
e-mail: kuzlyakina@msun.ru

J. C. García-Prada and C. Castejón (eds.), *New Trends in Educational Activity in the Field of Mechanism and Machine Theory*, Mechanisms and Machine Science 19,
DOI: 10.1007/978-3-319-01836-2_9, © Springer International Publishing Switzerland 2014

to the search activity and intends to use modern technologies and training methods which is based on new principles and techniques that are closely associated with the computer technologies. Nowadays training on course "Theory of Mechanisms and Machines" is intended to increase the natural sciences base of training, to introduce to the educational process mainly analytical methods of research and design of objects, to use the advanced and professional system for computer-aided design, and the availability of modern laboratory facilities and computer equipment. Computer technologies of training can change the forms and methods of training fundamentally, including the modern way of information's present, methods of amplification of person's creativity. Nowadays the task of creating of information environment on course "Theory of Mechanisms and Machines" on base of computer-aided training system is especially important.

Information environment which created by modern means of information technologies is considered as an integral part of the training environment and formed as an integrated system, whose components responded to academic, extracurricular, scientific research, checkout and training estimation.

2 The Structure of Information Environment

The structure of information environment is represented as interdependent set of software modules that provides possibility of arrangement and hold an educational process and realization of the functional responsibilities of users of any categories, the main ones are teacher and student. The composition and content of information resources is determined by teacher and a set of services—by software of computer-aided training organization system (CTOS). The basis for the organization of information environment on the course of TMM in MSUN named after adm. G. I. Nevelskoy for more than 15 years is CTOS COBRA [1]. The main content of information environment which regarding to training process are Theory, Practice, Calculations, Testing. Besides there are additional content for organization of information environment: user logging (for students), checkout of test's result by student, accessing of calculator, etc. Each content has own structure which lets to activate different information resources.

Information environment on course "Theory of Mechanisms and Machines" (IE TMM) includes:

1. Fragments of computer textbooks (two items)
2. Lecture notes (seven modules, there are some lectures in each module, there are 25 lectures in PPS formats)
3. Laboratory course with computer support (19 items in PPS, HTML, PDF, EXE formats)
4. Test check (input and output check includes five types of tests)
5. Calculation package for course designing and individual work (five packages)
6. Demonstration materials (more than 40 items in different formats)
7. Textbooks with different handles (six items in HTML, PDF formats)

Development and implementation of the information environment on the course TMM require careful approach and understanding of importance of using of each its elements. Filling the information environment allows providing information in various formats, and structured according to several of specialties. Mobility of information environment allowing make adjustments quickly, attach additional calculation packages and other materials, move and create new options and export them to various subsystems and to different users.

3 Information Resources of Environment on Course TMM

Discipline TMM is first engineering course in training structure of mechanic engineer in which such a subject as mathematics, physics, perspective geometry, theoretical mechanics, strength of materials, materials engineering, engineering graphics are integrated. A large amount of information, a high level of difficulty of content and different forms of training sessions forced to make a training process on modular approach. Module is a logically completed part of the course which includes: input and output control (testing), lectures, laboratory works, practices, course design.

Modular approach of training process is defined not only by the complexity and volume of material, but by modern trend to the wide use of computer technologies. Let's consider the main contents of information environment.

3.1 Theory

There are computer books, lecture notes, demonstration materials and other resources are represented in the Theory. Series of lectures is more rational to represent in pps format, because Power Point has a lot of functions for visual presentation of text (fonts, colour, animation, etc.), and has a possibility using hyperlinks.

As part of each module (Fig. 1) there is a file with incomplete drawings and diagrams on the subject module, which are used in time of giving classes in the classroom and in time of the self-tuition.

3.2 Practice

The most important element of any form of training is practice. IE_TMM includes three types of tutorial (Fig. 2): laboratory course, independent work, instructional design (course design, term paper, calculation and diagram design).

The greatest value in the training of engineers has laboratory course. The basis of laboratory work is practical examples which analyzing in the theory from lectures. It integrates the theoretical and methodological knowledge and practical skills of

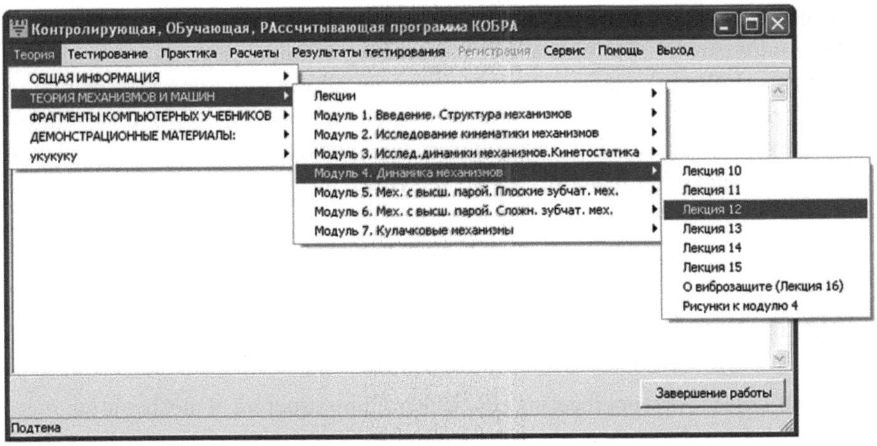

Fig. 1 The structure of information presenting (theory)

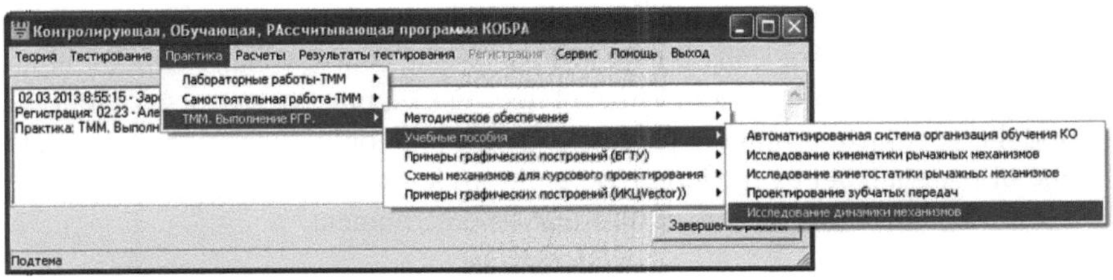

Fig. 2 The structure of information presenting (practice)

students within the framework of a unified process of work of educational and research character.

Laboratory course with computer support is performed on laboratory benches and on personal computers (PC) in program environments VSE, DINAMIC, GCG & FQ, and APM Win Machine [2]. The procedure of performing of laboratory works consists of six stages:

- Theoretical training—recommended practices to all laboratory work in text format in the library, in the Curriculum Office, and also it's embedded in hypertext format in CTOS COBRA and presented in the computer network of the university;
- Performing the necessary preliminary calculations and check their results on personal computer;
- Performing experiment on laboratory benches;
- Simulation on PC;
- Execution of report;
- Testing on subject of laboratory work in "COBRA" system.

Laboratory course consist recommended practices for working in special laboratory equipment and simulation of calculation procedures on a computer (Fig. 3).

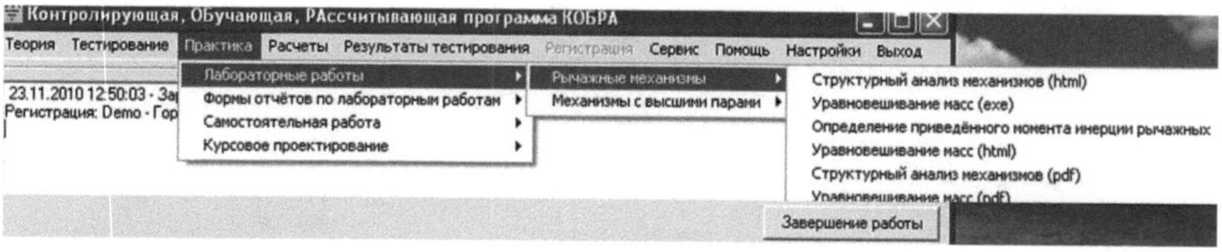

Fig. 3 Fragment of information environment. Practice. Laboratory works

3.3 Calculations (Course Design)

While studying course TMM in the process of performing of curriculum engineering calculations within a framework of course design is performed in the first time. While performing course design student activity is enshrined: work on the model, or search of rational solutions to accomplish the task at hand. For the first case there are examples in the information environment: the construction of mechanisms scheme, schemes of analog of the velocity and analog of the acceleration, forceful calculation and construction of forces schemes. Examples of construction are presented in presentations and step by step show the procedure of performing of graphical part of course design. In the second case special packages are used ПрИМеНЯюТсЯ сПеЦИаЛЬНЫе ПaketЬI: VSE, GCG&FQ, DINAMIC, APM WinMachine, etc., for graphic construction—packages: AutoCAD, COMPAS [3].

Tasks on course design must satisfy future profession of students. For the preparation of mechanic engineer of maritime fleet as tasks are slider-crank mechanisms which are in the base of units of marine equipment: main ship engine, diesel generating set, piston pump, blowers and others. For students on specialty Service of handling equipment of ports and transport terminals mechanisms of conveyor drive are offered. The screen of presenting of information resources on Calculations is on Fig. 4.

Design of plant on stage of structural and parametric analysis and synthesis in the training design performs by graphic and analytical methods in VSE, DINAMIC, GCG&FQ, Win Cam APM WinMachine systems. The screens of performing of parts of course project on theme “Research and design of plant and equipment. Mechanism of rocking conveyor” are shown on the Figs. 5, 6, 7, 8, 9, 10. Kinematic scheme of the mechanism is shown on Fig. 5; the dissection of the structural groups is on Fig. 6.

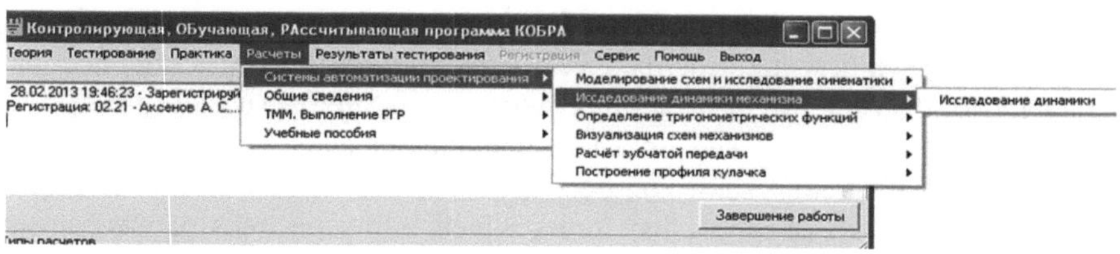

Fig. 4 Fragment of information environment. Calculations

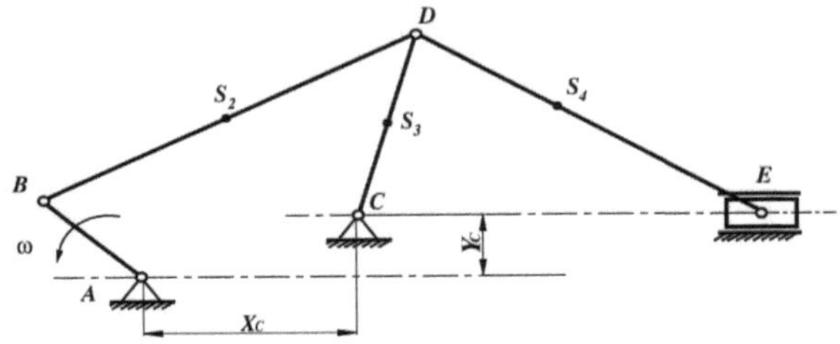

Fig. 5 Type diagram

Fig. 6 Assur groups of mechanism

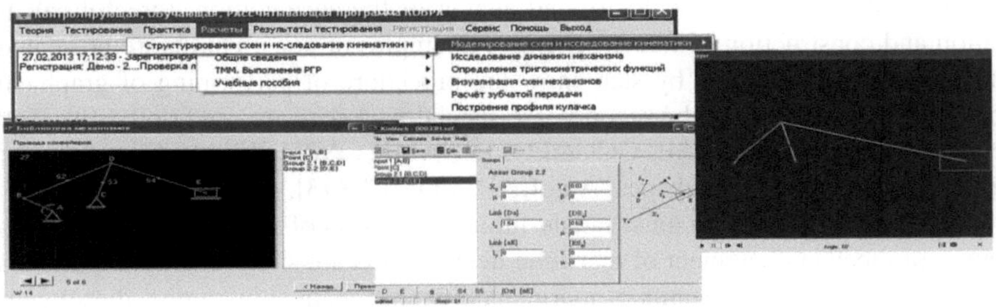

Fig. 7 Diagram of structuring and visualization of mechanism

Fig. 8 Dynamic model

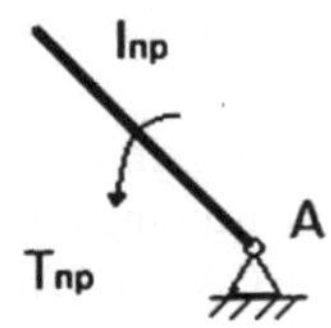

In the base of VSE package is simulation of mechanism diagrams and kinematics research within the generalized structural modules which are described earlier in papers [4–6]. Generalized structural modules allow even at the stage of structural and parametric synthesis to prescribe constructive form of links by introducing the parameters of the center of mass of links. Assur groups are the special case of generalized structural modules. The procedure of simulation and dynamic visualization of mechanism's diagram is shown on Fig. 7.

On the initial step of the design forces which acting on links of mechanism are unknown that's why necessary to determine the geometrical analogs of kinematic parameters which depend from the structure and geometry of a mechanism.

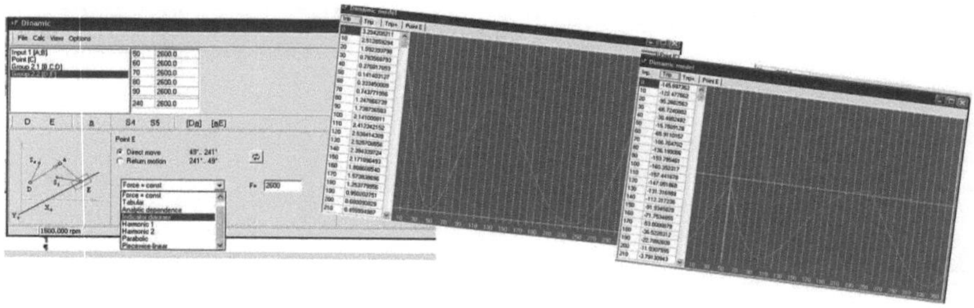

Fig. 9 The screen of mass entry, moments of links inertia, simulation of process impact and graphs of dynamic model's parameters

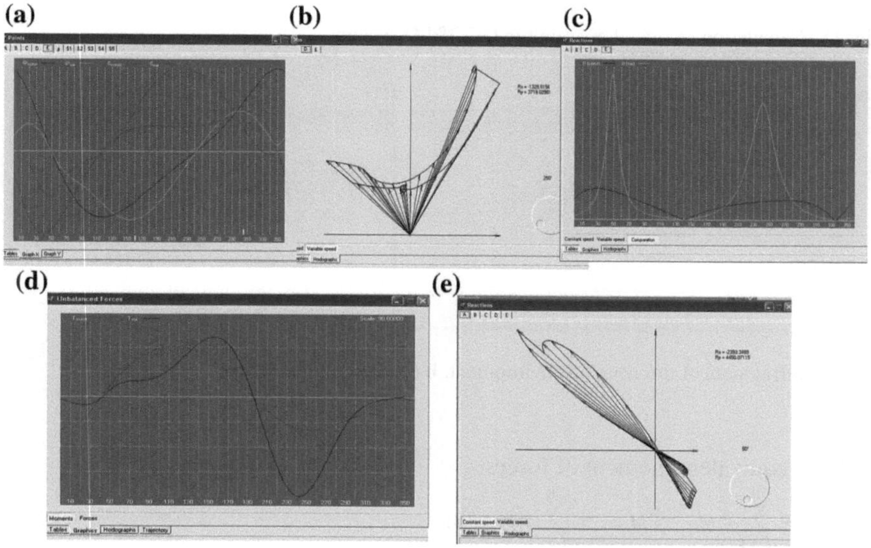

Fig. 10 Dynamic research within constant and variable velocity of input link: **a** time diagram of point E, **b** locus of reaction in joint D, **c** graphs of reaction in sliding pair E, **d** graphs of unbalanced moment of inertia, **e** locus of reaction in joint A

The geometric analogues of kinematic parameters of the main points and links are determined in the VSE system. Results of research display as tablets, graphs and locus diagrams.

An important task is to research mechanism's dynamic. On the initial step simple dynamic model—distributed mass which pivots around fixed point is assumed (Fig. 8).

Parameters of simple dynamic model:

I_{np}—equivalent moment of inertia

$$I_{\text{ïð}_j} = \sum_{i=1}^{n} m_i \cdot (V_{S_i\varphi_j})^2 + \sum_{i=1}^{n} I_{S_i} \cdot (\omega_{i\varphi_j})^2$$

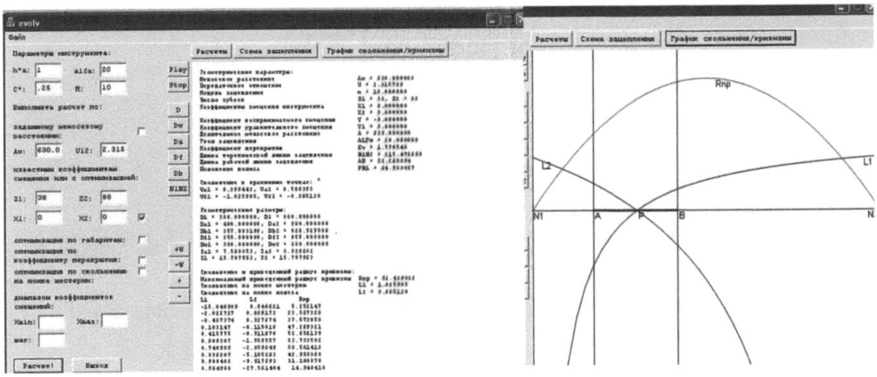

Fig. 11 Calculation results of gear unit in GCG&FQ system

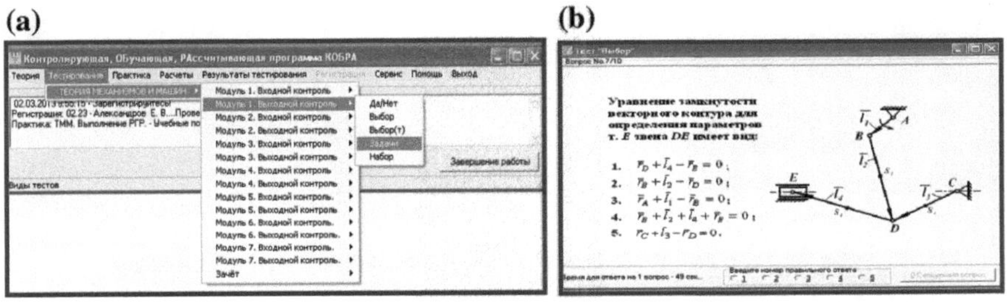

Fig. 12 **a** fragment of information environment, **b** example of test "choice"

T_{np}—equivalent moment of forces

$$T_{\ddot{\imath}\eth_j} = \sum_{i=1}^{n} \bar{F}_{M_{ij}} \cdot \bar{V}_{M_i\varphi_j} + \sum_{i=1}^{n} T_{i_j} \cdot \omega_{i\varphi_j}$$

The screen of parameters entry for determines the parameters of dynamic model and graphs of calculation results are shown on Fig. 9—$I_{\pi p}$ И $T_{\pi p}$.

On the initial step of design velocity of input link was constant. In the DINAMIC system equation of motion in machines is solved and motion law of input link, the main points and links and effective forces are determined. Then moment of inertia of flywheel to regulate the velocity fluctuation of input link in prescribed limits, unbalanced forces of inertia and force calculation are determined [3, 7, 8]. Some results of dynamic research within constant and variable velocity of input link are shown on Fig. 10.

In training (course) design plant design is performed. After dynamic research electric motor is selected from catalog and calculation of several flavors of gear unit: zero, positive and negative is made, compare of parameters and quality index is performed.

Use of professional packages in course design is helpful to learn design skills more effectively and its goal to help future specialists to value the structure impact of mechanism on their technical features. The experience which is taking within using VSE, DINAMIC, ADM WinMachine and other packages can be used in design project on special disciplines and in graduation course.

An important element of the information environment is textbooks in electronic form (Fig. 2). There are theory and practice that allow performing the course design on each part independently fully in there (Fig. 11).

3.4 Testing

A test system which includes five types of tests is particularly interesting (Fig. 12).

Test system is detailed in paper [9].

4 Conclusion

Modern technologies and means of training on course TMM can organize collective work in the classroom. The principle of filling the information environment which developed by IT-technologies, provides a consistent process of creation of information resources and for training control. Educational information environment can effectively coordinate the set of work while training in order to substantially improve the quality of professional education using new information technologies. It's helpful to expansion of cooperation in teaching staff, activization of scientific activities of all participants in the educational process. Information environment can be installed in a network version in the classrooms, as well as record on the disk for using on local computer.

Information environment on course TMM which presented in the paper is the result of years of research in automation of engineering design and education КОТОрЬIе ПреДсТаВЛЯЛИсЬ На МНОГИх МежДуНарОДНЬIх КОНФереНЦИЯх, сЦМПОЗЦуМах Ц ФруМах. About 200 papers are published, there are some titles in the references.

References

1. Kuzlyakina, V.V., Nagaeva, M.V.: Computer-aided training organization system ≪COBRA≫, p. 35. MSUN, Vladivostok (2010)
2. Nagaeva, M.V.: Laboratory course with computer support. Theor. Mech. Mach. J. No.1. **10**(19), 53–64 (2012) 10—St.Ptb., SPSTU press 2012
3. Kuzlyakina, V.V., Nagaeva, M.V.: Research of mechanisms kinematics and dynamics using CAD, p. 104. MSUN, Vladivostok (2010)
4. Kuzlyakina, V.V.: Lever mechanism computer-aided design system. In: Proceedings of the 11th World Congress in Mechanism and Machine Science, IFToMM 11, p. 5. Tianjin (2004)
5. Kuzlyakina, V.V., Slepenko, U.N.: Articulated lever mechanisms schemes structuring on the basis of generalized structural modules. In: Proceedings of the Nineth IFToMM International Symposium on Theory of Machines and Mechanisms, p. 10. Bucharest (2005)

6. Kuzlyakina, V.V., Nagaeva, M.V., Brazhnik, L.A.: Structuring of linkage diagrams (computer-aided system "VSE"), p. 76. MSUN, Vladivostok (2009)
7. Kuzlyakina, V.V., Brazhnik, L.A.: Definition of lever mechanisms dynamic model parameters at the initial stage of design. In: Proceedings of the 11th World Congress in Mechanism and Machine Science, IFToMM, p. 6. Besançon (2008)
8. Kuzlyakina, V.V., Slepenko, U.N.: Forceful calculation of linkage on the basis of generalized structural modules. In: The First IFToMM Asian Conference on Mechanism and Machine Science, p. 6. Taipei (2010)
9. Kuzlyakina, V.V., Holina, N.N.: Particularities to extracurricular work organization when studying engineering disciplines. In: Materials of Symposium IGIP 37 Engineering Competencies—Traditions and Innovations, p. 67–75. Moscow (2008)

An Example of Inquiry-Based Learning for Undergraduate Mechanical Vibrations

J. M. Chicharro, A. L. Morales, A. J. Nieto and P. Pintado

Abstract Laboratory or practical work is considered a crucial learning component for students of science and technology. It will be argued in this paper that the trend for inquiry-based learning that is beginning to percolate through classroom teaching can be brought to the design of laboratory tests. A project based on the free vibration of a one-degree-of-freedom pendulum serves as an example to show the integration of guided inquiry-based learning in a mechanical engineering curriculum. The goal is to design a pendulum system with a certain frequency of vibration and the challenge is to do it with specific elements and restrictions which, as the student needs to find out, do not lead to a solution just based on selecting a suitable pendulum length. In this laboratory teaching session, students were required to: identify the problem, model the system, derive governing equations, simulate, design, build, test, evaluate and share the solution. They are also required to experimentally measure the natural frequency and damping ratio of the system they built and adjust models and simulations accordingly. On the other hand, the session involves, albeit simplified, many of the common activities that the practicing mechanical engineer undertakes: design, develop, manufacture, assemble and test mechanical devices, including tools, engines, and machines.

Keywords Mechanical vibrations · Engineering education · Inquiry learning · Design process

1 Introduction

Laboratory experiments and practical work is considered a crucial learning component for students of mechanical engineering. The cost and time required to set up laboratory experiments is a drawback. Nevertheless, in the last years, mechanical

J. M. Chicharro (✉) · A. L. Morales · A. J. Nieto · P. Pintado
E.T.S.I. Industriales, Universidad de Castilla-La Mancha, Ciudad Real, Spain
e-mail: josemanuel.chicharro@uclm.es

J. C. García-Prada and C. Castejón (eds.), *New Trends in Educational Activity in the Field of Mechanism and Machine Theory*, Mechanisms and Machine Science 19,
DOI: 10.1007/978-3-319-01836-2_10,

vibrations laboratories have started to use simulations for the study of vibrations in mechanical systems [1], systems based upon virtual instruments [2], and even remote laboratories for self-learning [3, 4] which replace the traditional work in real laboratories.

However, many graduated engineering students have the wrong perception that experiments are only useful to write down data [5]. This vision of the students is motivated by the difficulty of carrying out in the laboratory all the engineering steps required to develop a certain product: design, develop, manufacture, assemble, and test mechanical devices, including tools, engines, and machines [6]. A solution to these problems seems to be found in "inquiry-based learning".

According to the American National Academy of Sciences [7] "scientific inquiry refers to the diverse ways in which scientists study the natural world and propose explanations based on the evidence derived from their work. Inquiry also refers to the activities of students in which they develop knowledge and understanding of scientific ideas, as well as an understanding of how scientists study the natural world". Then, the first question to ask when determining whether an activity is inquiry based is, "are students answering a research question through data analysis?" [8]. Banchi and Bell [9] suggested that there are four levels of inquiry-based learning in science education: confirmation inquiry, structured inquiry, guided inquiry and open inquiry. The level depends on the amount of information provided to students (e.g., questions, procedures and expected results) and on the guidance provided by the instructor.

In this work, a project based on the free vibration of a one-degree-of-freedom pendulum will serve as an example to show the integration of guided inquiry-based learning in a mechanical engineering curriculum. In particular, we propose to use the more complex methodology (open inquiry), where students must face a question and, given a list of materials, research, formulate, design and select their own procedures.

An additional goal is to integrate in the same activity several software tools for simulation, design and test which are commonly used by the students in the curricula corresponding to the degree in Mechanical Engineering of the Higher Technical School of Industrial Engineering of Ciudad Real (UCLM).

This curricula needs to be explained in the current context. During the 1990s, all Spanish universities updated the syllabuses of their courses as a result of the enactment of the new Organic Law of Universities. As a result, and for the first time, Computer Aided Design (CAD) appears in the list of core subjects in many of the technical degree courses [10]. Nowadays, these tools are integrated in all the academic years due to the massive use of these tools in industry and their benefits, such as [11]:

- Students can be provided with a more real-world experience.
- The problems which are assigned to students can be more complex.
- A solid foundation in computer aided design/analysis/manufacturing within the mechanical engineering curriculum will satisfy the needs and desires of both industry and students.
- Faculties will be able to develop the analytical and/or design skills necessary to carry out applied research.

Later, the entrance of Spain in the European Higher Education Area by the Bologna Declaration motivated the creation of new degrees which were started in 2010. In particular, the Higher Technical School of Industrial Engineering of Ciudad Real transformed the old studies of Industrial Engineering (5 years) into a 4 year Mechanical Engineering degree. During this process, the Department of Mechanical Engineering decided to update the contents of its courses.

The laboratory project proposed in this work corresponds to the updated course on "Mechanical Vibrations" (reference 56.374), which is offered to fourth year students. One specific competence of this subject is that the student must know the analysis and measurement techniques of vibrations in machines and structures, along with the capacity to study stresses and strains via finite element methods.

2 Learning Methodology

A project based on the free vibration of a one-degree-of-freedom pendulum serves as an example to show the integration of guided inquiry-based learning in a mechanical engineering curriculum. The goal is to design a pendulum system with a certain frequency of vibration and the challenge is to do it with specific elements and restrictions which, as the student needs to find out, do not lead to a solution just based on selecting a suitable pendulum length.

In this case an open inquiry methodology is employed (level of inquiry 4): the student must investigate topic-related questions and formulate solutions through self-designed/selected procedures. The inquiry matrix will serve as the guide to implement science inquiry along the teacher-explicit to student-initiated continuum (see Table 1) [12].

The challenge starts formulating a very simple question to groups of 2 or 3 students: "Design and build a mechanical system with a constant vibration frequency of $f_n = 1.4$ Hz, only by using the given materials; then, measure the damping and adjust the dynamic model". A list of elements is provided for the manufacturing process of the simple pendulum: a support, a bearing, bars of lengths from 150 mm to 200 mm, disc-shaped masses of diameters from 80 mm to 120 mm and a steel wire roll of 0.4 mm in diameter.

Table 1 A matrix of open-inquiry science learning

Open-inquiry components	Pre-lab (stage 1–2)	Laboratory (stage 3)	Post-lab (stage 4)
Teacher	Open-ended problem/question	Gather the materials	
Student	Background/theory	Built	Draw a conclusion
	Make a hypothesis	Test and evaluate	Reporting
	Simulate	Redesign	
	Design		

Fig. 1 Design process

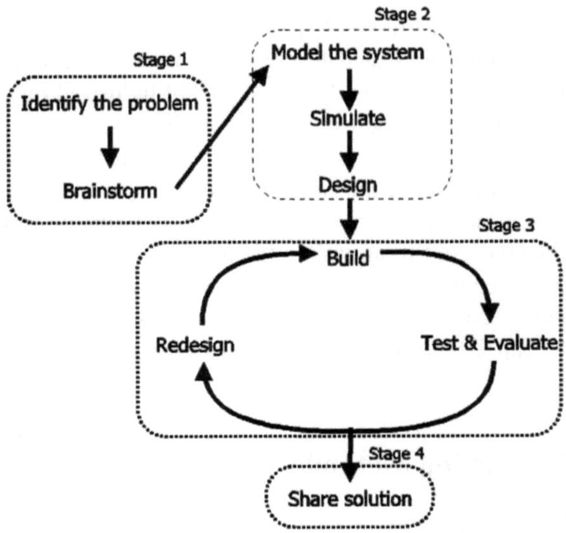

Figure 1 shows the design process which is proposed to students. It is divided into four stages which are carried out, as shown in Table 1, before the laboratory work (stages 1–2), in the laboratory (stage 3) and after the laboratory work (stage 4). Note that when an engineer is solving a problem rarely the solution arises from the first ideas, so it is not the aim of this methodology to propose easy problems. In the example we are presenting, the basic idea of changing the length of the simple-pendulum to fit the natural frequency does not lead to a successful solution.

During stage 1 the student must think a solution and can only gather information related to the topic. Obviously, she or he will need to build a simple-pendulum with the materials provided and get the ordinary differential equation which governs the motion and the natural frequency of vibration:

$$L_G^2\ddot{\theta} + gL_G \sin\theta = 0 \quad \rightarrow \quad f_n = \frac{1}{2\pi}\sqrt{\frac{g}{L_G}} \tag{1}$$

where m is the mass, L_G is the distance between the centre of mass and the support, g is the gravity and θ is the angle.

At this stage, the student must realize that this solution is not valid. In fact, even considering the shortest bar ($L_B = 150$ mm) and the disc with the smallest radius ($R = 40$ mm), $L_G = L_B + R = 190$ mm, which implies a natural frequency of $f_n = 1.15$ Hz, 18 % lower than the goal. Since no shorter length is available, the only option is to include additional stiffness (leaving aside the far-fetched solution of modifying gravity).

If the system is modeled without neglecting the inertia (I_G) and damping (C), and including the stiffness of a torsion spring (K), the equation of motions results:

$$\left(mL_G^2 + I_G\right)\ddot{\theta} + C\dot{\theta} + (mgL_G + K)\,\theta = 0 \tag{2}$$

The damped frequency of vibration f_d which results from Eq. (2), considering that the system is underdamped, is:

$$f_d = \frac{\sqrt{1-\xi^2}}{2\pi}\sqrt{\frac{K + mgL_G}{mL_G^2 + I_G}} \tag{3}$$

where ξ is the damping coefficient, i.e. the relation between the damping and the critical damping of the system [13].

The required increase of stiffness K can be achieved by means of manufacturing a helical torsion spring with the steel wire roll. Such stiffness is given by [14]:

$$K = \frac{Ed^4}{DN_a} \tag{4}$$

where E is the Young's modulus of the material, d is the diameter of the wire, D is the diameter of the roll and N_a is the number of equivalent turns which the torsion spring has.

Once the decision about the proposed solution has been made, the student gets into stage 2, where he or she must simulate the results and design the mechanism. First, computational software will be used to obtain the temporal response of the ODE and its FFT in order to see the effect of including the torsion spring in the frequency domain. Figure 2 shows such spectrum, which compares the simple pendulum with the smallest bar and disc and the pendulum along with the torsion spring (roll diameter of 15 mm and 8 turns). The final design is obtained via CAD/CAE software, where the response can be verified.

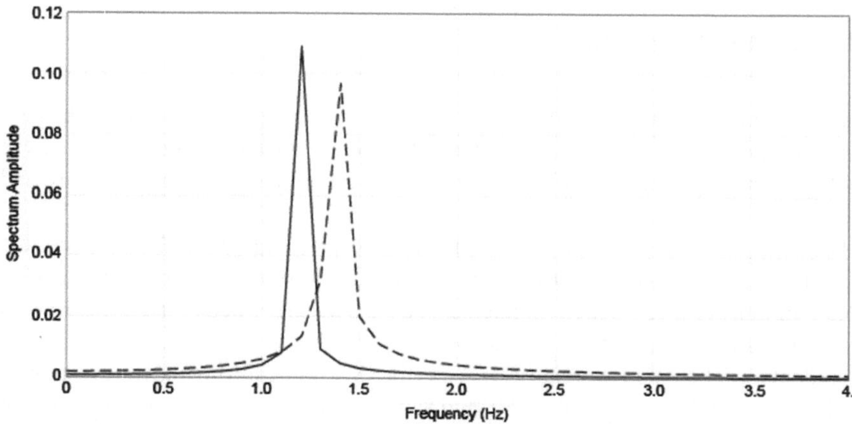

Fig. 2 Simple pendulum (*solid line*) and pendulum with helical torsion spring (*dotted line*)

Stage 3 comprises the manufacture of the pendulum with the helical torsion spring. In addition, students are also required to experimentally measure the natural frequency and damping ratio of the system they have built and adjust models and simulations accordingly. They can use, for instance, the logarithmic decrement method [13]:

$$\delta = \frac{1}{n} \ln \left(\frac{X_1}{X_n} \right) = 2\pi\xi \tag{5}$$

Once the proposed solution has been evaluated and its results are successful, the student gets to stage 4, where he must report his work and share the solution.

During the challenge, the student is told to use the software which has learnt to use in his courses in an order which simulates the real design of a product in the industry:

- *MatLab*, which will be used to obtain the temporal response of the schematic mechanical system. It is a numerical computing environment developed by MathWorks that allows matrix manipulations, plotting of functions and data, implementation of algorithms, creation of user interfaces, and interfacing with programs written in other languages…
- *SolidWorks*, for the design of the solution which fits the requirements of the challenge and for verifying the response with its simulation utilities. It is a 3D mechanical CAD program that provides a suite of product development tools for mechanical design, design verification, data management, and communication tools.
- *LabView*, for testing the prototype, registering data and analyzing the response in the time and frequency domains. This software represents a programming environment that includes specific tools for instrument control, data acquisition, storage, analysis, presentation and integrating of those features in a single system. It uses graphical programming and a special graphical language.

3 Experimental Setup

Stage 3 is carried out in the laboratory, so it needs to be properly suited with the necessary equipments. In particular, students will need to register the signal and analyze it in the frequency domain, which is achieved by using properly some software and hardware.

The devices that are used for data acquisition are a PC with PCMCIA connectivity, a National Instruments DAQCard-AI-16E-4, a SCC-2345 carrier equipped with SCC-ACC01 modules for ICP accelerometers and a PCB accelerometer type 333B50 with a frequency range from 0.5 Hz to 3,000 Hz.

Students will be asked to create a *LabView* file (VI) for acquiring the time response, filtering the signal, obtaining the FFT and saving results in a file. Figure 3 shows an example of VI with all the required capabilities.

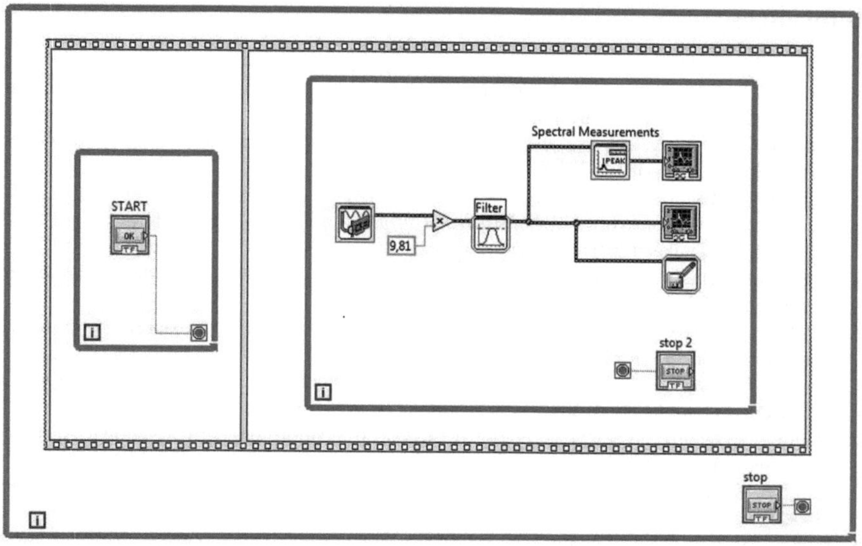

Fig. 3 *LabView* file for the measurement of pendulum acceleration

4 Conclusions

To integrate all the practical activities that a mechanical engineer will develop during his or her career in just one laboratory work may be very complicated unless this work corresponds to his undergraduate dissertation. This article makes an effort to overcome these difficulties and shows an example of practical project which includes all the steps of the industrial design process, to be carried out during a course of Mechanical Vibrations.

The project has been designed to require the use of all the software utilities which the students use in their Mechanical Engineering studies. In this way, the student can contextualize the use of these tools within the different stages of a full product design. The concept that laboratory work is a key stage of industrial design is emphasized.

The pedagogic methodology that has been used is inquiry-based learning, which enhances the collaborative competences and highlights the roll of the student as an active subject in charge of his learning. Scientific competences are also strengthened since students are faced up to a problem that they need to solve by means of their scientific knowledge.

The development of all these competences is very important for the integration of the engineering students in industry or scientific research.

Acknowledgments The authors are grateful for the support received from the project "Convocatoria 2012 de Ayudas Contrato Programa para material fungible de prácticas de laboratorio" financed by the Higher Technical School of Industrial Engineering of Ciudad Real (University of Castilla-La Mancha).

References

1. Wei-Pin, L., Ming-Shiun, L.: A digital simulation of the vibration of a two-mass two-spring system. Comput. Appl. Eng. Educ. **18**, 564–573 (2010)
2. Gillich, G.-R., et al.: The use of virtual instruments in engineering education, Procedia Soc. Behav. Sci. 2, 3806–3810 (2010)
3. Aziz, E.-S., Esche, S., Chassapis, C.: On the design of a virtual learning environment for mechanical vibrations. In: 37th ASEE/IEEE Frontiers in Education Conference, vol. F2H, pp. 7–12. Milwaukee (2010)
4. Elgamal, A., et al.: Webshaker: live internet shake-table experiment for education and research. Comput. Appl. Eng. Educ. **13**, 99–110 (2005)
5. Bilal, N., et al.: Reversing the roles of experiment and theory in a roving laboratory for undergraduate students in mechanical vibrations. Int. J. Eng. Educ. **21–1**, 166–177 (2005)
6. Bureau of Labor Statistics: Occupational Outlook Handbook, 13th edn. Mechanical Engineers, U.S. Department of Labor (2012)
7. National Academy of Sciences: National Science Education Standards. National Academy Press, Washington (1996)
8. Bell, R., Smetana, L., Binns, I.: Simplifying inquiry instruction: National science teachers association. Sci. Teach. **72**, 30–33 (2005)
9. Banchi, H., Bell, R.: The many levels of inquiry. Science and Children **46**(2), 26–29 (2008)
10. Rubio, R., et al.: Present state of CAD teaching in Spanish Universities. Comput. Educ. **44**, 201–215 (2005)
11. Sorby, S.A., et al.: Modernization of the mechanical engineerring curriculum and guidelines for computer-aided engineering instruction. Comput. Appl. Eng. Educ. **7**, 252–260 (1999)
12. Cuevas, P., Lee, O., Hart, J., Deaktor, R.: Improving science inquiry with elementary students of diverse backgrounds. J. Res. Sci. Teach. **42–1**, 337–357 (2005)
13. Rao, S.(ed.): Mechanical Vibrations, 5th edn., Pearson, London (2010)
14. Hamrock, B., Jacobson, B., Schmid, S.: Elementos de máquinas. Mc GrawHill, Mexico (2000)

Application of Techniques Based on Reliability for the Improvement of the Teaching of Mechanisms in Industrial Design

Justo García Sanz-Calcedo, D.Rodríguez Salgado, I.Cambero and J.M.Herrera

Abstract The aim of this study is to analyse the importance of promoting the teaching of reliability of mechanisms in teaching degree in industrial design. An assessment is made of the need for these features provided from the earliest stages of product design, as this allows for a higher quality and economy. The industrial design aims to create a manufacture, device, machine or useful system that is efficient, safe and practical. This is only possible if the correct definition of the functions and design specifications, where an important aspect is the reliability of the designed object, something that should be thoroughly detailed in the conceptual design stage. The experiment was carried out with students in the degree of Industrial Design and Product Development at the University of Extremadura (Spain).

Keywords Technology · Earning · Reliability · Design · Maintainability

1 Introduction

Product design is concerned with designing capital equipment and consumer, making it easier for users to understand how it works and how is it culturally accepted. The products design is a project activity, technological and creative [1].

To design properly a product, it is essential to take into consideration the reliability of each of the components that compound the product, to produce higher quality designs [2, 3].

Justo García Sanz-Calcedo (✉)
Design Engineering School, University of Extremadura, Badajoz, Spain
e-mail: jgsanz@unex.es

D. R. Salgado, I. Cambero, J. M. Herrera
Industrial Engineering School, University of Extremadura, Badajoz, Spain
e-mail: drs@unex.es

J. C. García-Prada and C. Castejón (eds.), *New Trends in Educational Activity in the Field of Mechanism and Machine Theory*, Mechanisms and Machine Science 19,
DOI: 10.1007/978-3-319-01836-2_11,

Nowadays the engineering design is performed by applying regulated by factors design codes. This approach ignores the random nature of the design variables [4] and is not flexible when considering the socioeconomic characteristics [5].

The reliability analysis comprises a set of mathematical and statistical methods, organizational procedures and operational practices that, through the study of the laws of occurrence of failures, are aimed at the resolution of the problems of prediction, estimation and optimization of the probability of survival, average length of life and the percentage of time in proper functioning of a system [6].

The reliability is a property intrinsic to a product, defined as the probability of meet a required function, in the conditions of use and for a certain period of time. There is no reliability without initial quality, because the reliability is an extension of the quality in the time [7].

The reliability-based design is an alternative to traditional design, in which the randomness of design variables and the use of uniform security measures at the time of the design of a product is considered.

The aim of this study is to analyse the importance of promoting the teaching of the reliability of mechanisms in teaching degree in industrial design [8]. An assessment is made of the need for these features are provided from the earliest stages of product design, as this allows a higher quality and economy. The industrial design aims to create a manufacture, device, machine or useful system that is efficient, safe and practical. This is only possible with the correct definition of the functions and product's design specifications. An important aspect is the reliability of the designed object, something that should be thoroughly detailed in the conceptual design stage [9].

2 Methodology

The experiment was carried out with students currently taking the Industrial Design and Product Development degree at the University of Extremadura (Spain), during the development of Technical Office's subject, in particular conducting a seminar-based on product design reliability [10]. In the seminar the students experienced a theoretical and practical perspective, the advantages and disadvantages of making the design of a product based on the same reliability [11], including a set of mathematical and statistical methods, organizational and operational procedures through the study of the laws of the occurrence of faults, which are used to obtain the probability of survival, mean length of life and the percentage of time of operation of a system [12].

The student's academic objectives consisted in introducing models, probabilistic and statistical methods used in the analysis of problems of reliability, in two stages: firstly, probabilistic models of time of a system of components, and secondly are examined and studied statistical methods that can be applied to a data set of survival times.

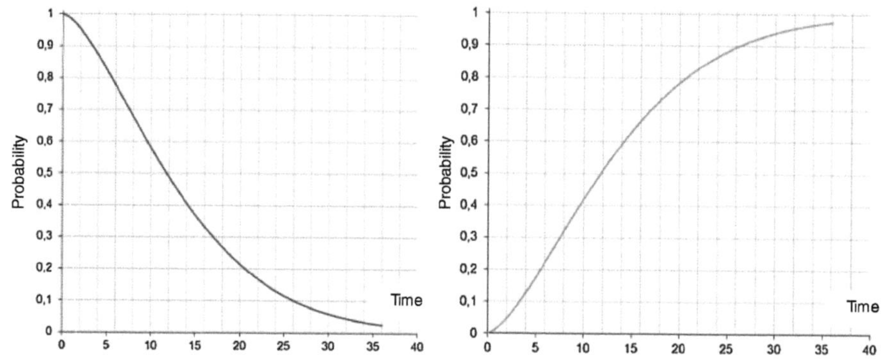

Fig. 1 Survival and fault functions

Other objectives were to provide a review of probabilistic and statistical common in reliability techniques, understanding the relationship between reliability and survival and promote the use of computer programs to solve problems of reliability.

To analyze the life of a component, the variable T is defined as the useful life of the asset or component, that is, the random variable that defines the concept of reliability, which is the time duration or life of the device.

It defines the survival function that gives the probability of a component that is working after *t* h. Thus, if a component has a function of R (10) = 0.57 means that the probability of the component working within 10 hours is 57. Figure 1 represents the survival function, best-known as a reliability feature.

On the other hand it defines the function failure, *F(t)* gives us the probability that a component fails after *t* hours. Thus, if a component has a function of F(20) = 0.89, means that the probability of component failure within 20 units of time is 89 %. Figure 1 shows a function corresponding to an electrical equipment failure.

Fault function is complementary to the survival function, so we can express to:

$$F(t) = 1 - R(t) \tag{1}$$

Figure 2 represents the survival function, hereinafter called function reliability and function failure, both complementary.

It is assumed that the random variable *T* is expressed by cumulative distribution function *F(t)* in mathematical form:

$$F(t) - P(T \leq t) \tag{2}$$

Then, the density function of failure, f (t), can be defined as:

$$f(t) = \frac{d}{dt}F(t) \tag{3}$$

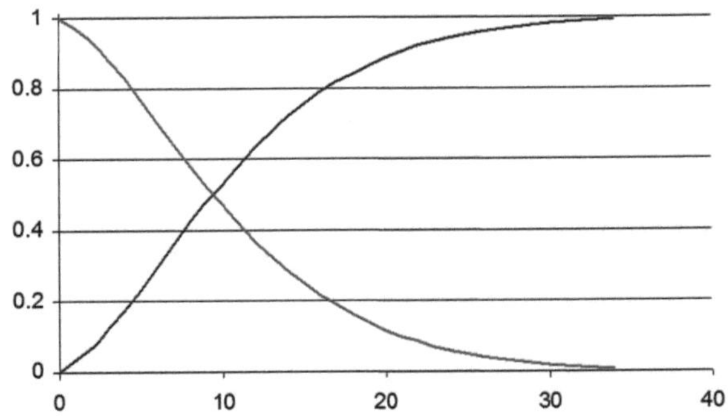

Fig. 2 Survival and fault functions

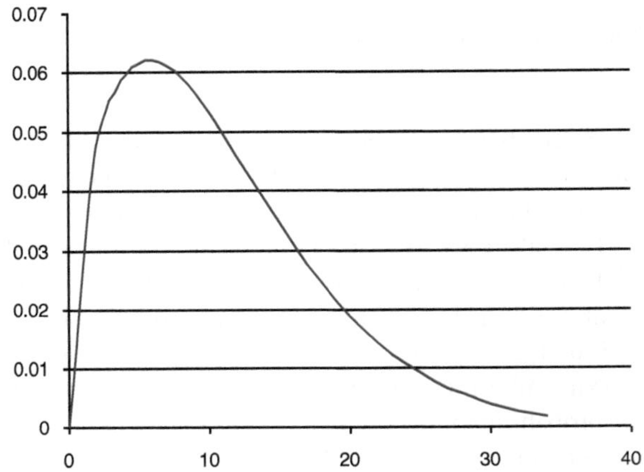

Fig. 3 Failure density function

In this way, we have a quantification of the probabilistic distribution of life dispersion. The Fig. 3 shows the graphical representation of the function density of electromechanical component failure.

This helps us to make decisions about the choice of a system or component. For example, it can be seen in Fig. 4, which flatter the blue function behavior, posed that failures have approximately the same chance of occur at any point in the range 1–20. However, the case of the function represented in red color, with a very targeted area, suggests the existence of a time interval. In this interval, the possibility of a failure from occurring is reduced. Therefore, technology corresponding to the first function is not allowed to set a time interval which can be assured with some confidence into the behavior of the manufacture, being the second function technology most suitable for this purpose.

In other words, *R(t)* is the probability that a new component could survive longer than t. Therefore, *F(t)* is the probability that a new component does not survive longer than t.

On the other hand, the probability that a new component fails between t and $t+s$, where s is an increase of time with respect to t, equals a:

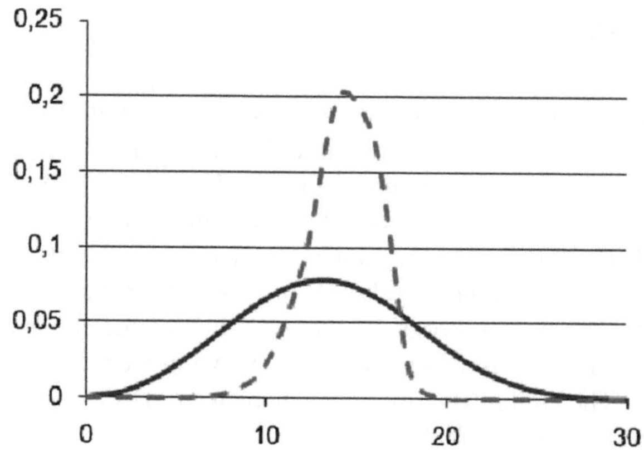

Fig. 4 Fault density curves of two manufactures

$$P(t < T) \leq (T + S \perp T > t) = \frac{P(t < T) \leq (t + s)}{P(T > t)} = \frac{F(t + s) - F(t)}{R(t)} \tag{4}$$

Dividing the fourth equation into s, taking limits and doing that s tends to zero:

$$\lambda(t) = \lim_{s \to 0} \frac{1}{s} \cdot \frac{F(t + s) - F(t)}{R(t)} = \frac{f(t)}{R(t)} \tag{5}$$

where $\lambda(t)$ is the function of failure rate or risk function or instant failure rate, and is a feature of a product's reliability. It represents a percentage of surviving devices in an instant t, and it can be defined as the relationship between the average number of faults and operating time.

The failure rate can be written based on reliability:

$$\lambda(t) = \frac{f(t)}{R(t)} = \frac{f(t)}{(1 - F(t))} \tag{6}$$

where $\lambda(t)$ is the failure rate, $R(t)$ is the reliability function and $f(t)$ is the probability density function being t the independent variable.

The failure rate can be interpreted as the "speed" in which faults could happen, it is a measure of the device that is prone to failure as a function of time which has been running.

The characteristic of reliability is the Mean Time Between Failure (MTBF). The MTBF corresponds to the expected value of the random variable t, time of occurrence of a failure, indicating an operating device, which one will inevitably suffer a failure at time t, that is a priori unknown.

The mathematical expression of MTBF, corresponds to the equation:

$$MTBF = \sum_{n=0}^{n=i} \frac{TBFi}{n} = \frac{1}{\lambda} \tag{7}$$

3 Results

The Weibull distribution allows us to study the distribution of a component failure and let us take control about faults vary over time. So we can determine an equation that relates us continuing failure probability with the time of outset.

The method does not determine which variables influence the failure rate, a task that will be left to the analyst, but facilitates the identification and consideration of those, being a powerful tool for predicting behavior. The Weibull distribution is so flexible by appropriately choosing the parameters, and allows to describe the three failure steps function rate curve shown in the bath-tub.

The Weibull distribution is characterized by two parameters: α, a (scale parameter) and β (shape parameter).

$$R(t) = e^{-\left(\frac{t}{\alpha}\right)^{\beta}} \tag{8}$$

Where *R(t)* is the reliability function, a parameter α associated with the characteristic lifetime of the equipment and ß is a parameter associated at the time. When the value of β is less than 1 indicates that the failure rate decreases with time and if the value is greater than 1, the failure rate increases with time. In case the unit, indicates that the failure rate is constant.

To calculate the value of α and β computer programs are based on the values of the failure times, calculating the parameters α and β. You can also use the methodology detailed in the NTP-331 reliability: the Weibull distribution [13], which uses the graphic resolution through a special role for graphics, called Weibull paper. The process chart requires several steps and one or two iterations, but it is relatively straightforward.

The life feature of manufactures is defined as the value that corresponds to a failure rate of 63.2 %, a value that is commonly used to determine the useful life of a computer.

4 Conclusions

To ensure the product quality, it is essential to be used in the design phase techniques based on the reliability of each component of a product. It is desirable in the descriptors of subjects related manufacturing processes, advanced knowledge in reliability, maintainability and availability.

Include in the subject of design concepts related to reliability, can provide students basic tools for the design and ensure quality product designed.

Acknowledgments The authors wish to express their acknowledgement to the Design Engineering School to the University of Extremadura (Spain), for their contributions to the development of this study. This work has been developed in the framework of the GR-10099 research project within the IV Regional Plan of I+D of Extremadura 2011–2014.

References

1. Feepd. El diseño en España. Estudio Estratégico. Federación Española de Entidades de Promoción de Diseño (2001)
2. Maldonado, T.: El diseño industrial reconsiderado. Colección punto y línea. Editorial gustavo gili, p. 13. S.A., Barcelona (1977)
3. Vandeville, P.: Gestión y control de la calidad. Ediciones AENOR, Madrid (1990)
4. Bonsiepe, G. Teoría y práctica del diseño industrial, Colección comunicación visual, E. (ed) Gustavo Gili, pp. 24–25. S.A. Barcelona (1978)
5. Northwood, M.D., et al.: Problem-based learning, From the health sciences to engineering to value—Added in the workplace. Glob. J. Eng. Educ. **7**(2):157–164 (2003)
6. Aguayo, F., Ramírez, A.: Ingeniería del Diseño y desarrollo de productos. E. (ed.) Tebar Flores. Madrid (1997)
7. Eisner, H.: Ingeniería de sistemas y gestión de proyectos. Ediciones AENOR (2000)
8. Hills, G., Tedford, D.: The Education of engineers: The uneasy relationship between engineering, science and technology. Glob. J. Eng. Educ. **7**(1), pp. 17–28 (2003)
9. Mcdermott, K.J., Göl, Ö., Nafalski, A.: Considerations on experience-based learning. Glob. J. Eng. Educ. **6**(1), pp. 71–78 (2002)
10. Aguayo, F., Soltero, V.M.: A. Metodología del diseño Industrial. Un enfoque desde la ingeniería recurrente E. (ed.) Ra-ma. Madrid (2002)
11. Andersen, H. Experiences from a pedagogical shift in engineering education. Glob. J. Eng Educ. **6**(2), pp. 139–144 (2002)
12. Monchy, F.: Teoría y Práctica del Mantenimiento Industrial. Masson, Ed. (1990)
13. Instituto Nacional De Seguridad E Higiene En El Trabajo. Fiabilidad: la distribución de Weibull. NTP 331

Part III
Mechanism and Machine Science in the Engineer Program: Applications and Research

Optomechatronics Applications of the Theory of Mechanisms with Active Student Involvement in Research

V.-F. Duma, A. Schitea, M. Tuef, O. Cira, C. Mnerie, Gh. Hutiu, D. Demian and I. Kaposta

Abstract The paper presents some of our current investigations in the multi-disciplinary field of optomechatronics, based in part on different applications of the theory of mechanisms. Classical mechanism applications approached mostly with undergraduate students are presented in the first part of the paper. Scanners, choppers and attenuators—optomechatronic devices, in general—are considered in the second part, with both kinematical and dynamical aspects, and some of our relevant results in the field are pointed out. Student involvement (both under and postgraduate) in these researches is presented, as well as some of the implementation of the results and expertise gained through research in the curricula of the Mechanical and Electrical Engineers in our university.

Keywords Higher education · Students · Research · Mechanisms · Mechatronics · Optomechatronics

1 Introduction

The 3OM Group (in Opto-Mechatronics, Optical Metrology, and Optics and Mechanics) we have established in 2008 [1] at the Aurel Vlaicu University of Arad (UAVA), Romania, has proposed from its beginning to translate the results and expertise of its research activity into the teaching of Undergraduate (UG) and into the training of Postgraduate (PG) students. We have presented in a previous study [2] a part of our methodology related to the effort of passing from classical teaching to hands-on-experience and to a research-related educational process—in conjunction with a most interesting experience as a Fulbright Fellow in a top-research American university and also with other studies on education [3].

V.-F. Duma (✉) · A. Schitea · M. Tuef · O. Cira · C. Mnerie · Gh. Hutiu · D. Demian · I. Kaposta
Aurel Vlaicu University of Arad, Arad, Romania
e-mail: duma.virgil@osamember.org

J. C. García-Prada and C. Castejón (eds.), *New Trends in Educational Activity in the Field of Mechanism and Machine Theory*, Mechanisms and Machine Science 19,
DOI: 10.1007/978-3-319-01836-2_12,

In the present paper we shall somehow continue this presentation, this time going into an overview regarding the specific topics we have approached in the last decade, with some relevant examples and with the experience of integrating the research expertise gained in different projects into the PG and even the UG students' curricula—both in Mechanical (ME) and Electrical Engineering (EE)—for the various subjects we teach. We shall mainly focus on topics related to the Theory of Mechanisms and to its sometimes surprising applications in inter-disciplinary fields, such as Optical Engineering or Optomechatronics (OME).

The research activities involving students in our group can be divided roughly in two (regardless of the fact that they are UGs or PGs): (i) research in classical areas, strictly limited to a specific topic, such as the Theory of Mechanisms—and some of these will be presented in Sect. 2; (ii) inter-disciplinary research, carried on in OME—with students that are willing to do this—and some of these will be presented in Sect. 3. The former direction involves of course a higher risk and usually a higher effort, but the results can be very rewarding—both for students and faculty.

2 Mechanisms Researches with Students in Our Group

A current activity in the Mechanisms Lab at UAVA—and in our group—is related to the construction of different mechanisms (linkages, gears and cams) [4], to their analysis and optimization. A few examples of mechanisms thus developed are presented in Figs. 1, 2 and 3.

The analyses of linkages [5, 6] have been performed with UG students using not only dedicated mechanism programs, such as SAM 6.1, but also general mechanical design programs that have the advantage to be used throughout the entire ME. Such an example is shown in Fig. 2, for a plane linkage developed with students (Fig. 2a) in the Mechanism Lab [7]. A simulation of its movement is performed both 2D (with CATIA V5R20 Sketch Design—Fig. 2b) and 3D (with CATIA V5R20 DMU Kinematics—Fig. 2c), and the movement laws obtained, including velocities and

(a) (b)

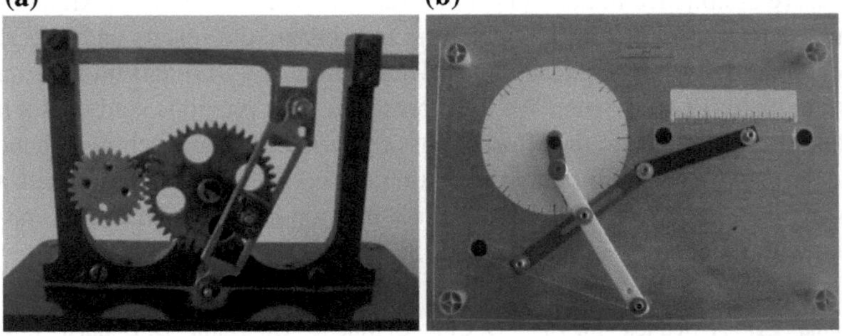

Fig. 1 Examples of linkages developed and studied with students in the mechanisms lab

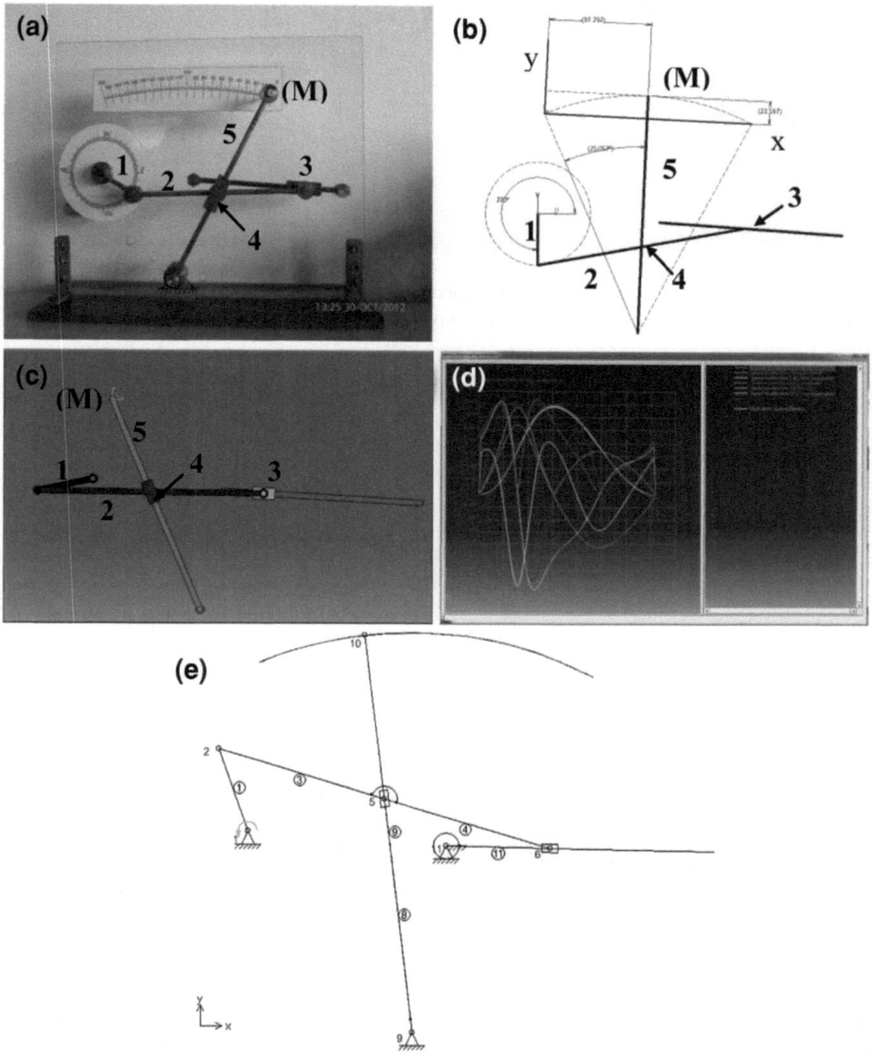

Fig. 2 Example of an analysis of a linkage—student project: **a** linkage constructed in the lab; **b** modeling with CATIA V5R20 Sketch Design; **c** modeling with V5R20 DMU Kinematics; **d** movement laws on x and y obtained with (**c**); **e** analysis—SAM 6.1 [7]

accelerations (Fig. 2d) confirm each other, and are verified with SAM 6.1 as well (Fig. 2e). Advantages and drawbacks of each method are derived. Students can remark that using CATIA one may delivers accurate results, but this involves a laborious process. However, with 3D CATIA, spatial mechanisms can be solved, while with SAM, only planar linkages can be approached.

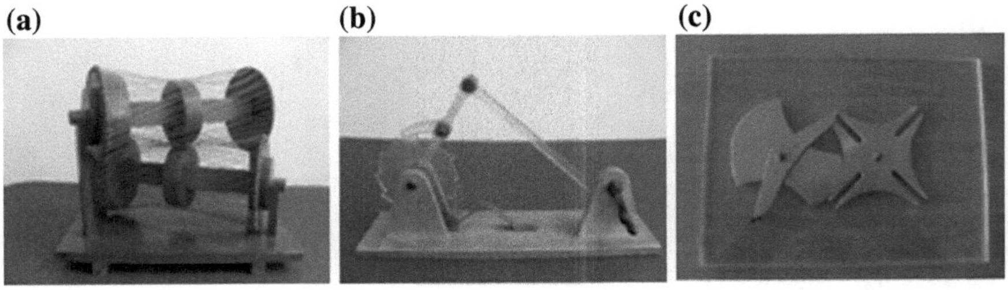

Fig. 3 Examples of gears constructed in the mechanisms lab at UAVA: **a** a general model of the hyperboloid gears; **b** a ratchet mechanism; **c** a type of Geneva mechanism

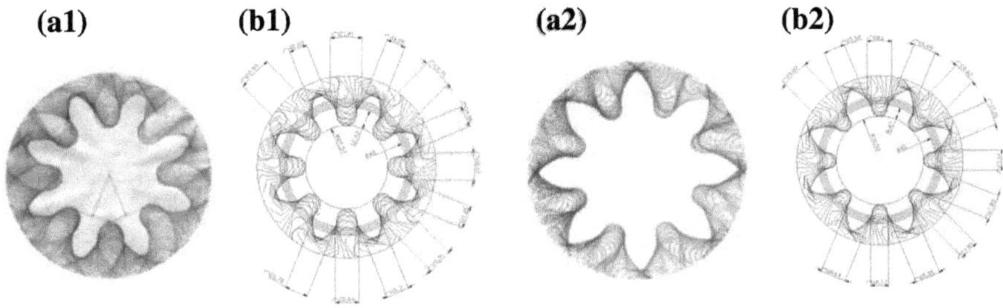

Fig. 4 **a** Experimental and **b** AutoCAD simulations of the gears manufacturing process. *Examples* gear with 10 mm module, 8 teeth, and profile movements: (*1*) 0; (*2*) 4 mm [8]

In Fig. 4 a part of a study we have made on the manufacturing process of gears [8] is presented. An AutoCad modeling was used (verified experimentally) to obtain profiles of the wheels with different parameters and profile movements.

Such studies were finalized for the students with a paper [4, 8, 9], a chapter in a book edited by faculty both elaborated with UG students [10], with communications at the students' scientific sessions (organized every year in different periods in each Romanian university), and eventually with a Diploma Thesis. A positive aspect is that since 2010 selected UGs (15 % of the total) can perform their mandatory summer internships in engineering in the university labs. The most passionate students may thus have a time dedicated entirely to an early research experience. In Mechanisms this happens from the 2nd year. The best and most motivated students may afterward (starting with the 3rd year) be selected as Research Assistants on projects—as presented in the next section. Their activity concluded with a Diploma work is usually carried on for their Master as well—with an output in terms of publications that can allow them to apply to pursuit Doctoral studies.

All the models developed, as well as the experience gained in Mechanisms is transferred to lectures and to laboratory works with the UG students. This also includes other topics except Mechanisms, as presented in Sect. 4.

3 Inter-Disciplinary Student Researches in Our Group

The directions of research in the 3OM Optomechatronics Group are divided in two: (i) Optomechatronics devices; (ii) Measuring systems, for Optical Metrology, Radiometry and Colorimetry, and latter Optical Coherence Tomography (OCT)—one of the most rapidly advancing techniques in biomedical imaging [11–13].

Students-involved research (especially UG) is usually focused on optomechatronic devices that are easier to approach and are more suitable to study in a diploma or dissertation thesis. These topics include for our Group:

(i) *Polygon mirror (PM) scanners* are the fastest optomechanical scanning systems [12, 13, 17–19]. We approached them as a rotating cam with the laser beam incident on it as a follower. We have worked on industrial measurements systems involving PMs (Fig. 5a) and we are currently working on PM filters for OCT [12, 13]. Kinematics studies [20] determined the scanning functions, i.e., the current position h of the beam—Fig. 5a, its velocity, the extreme positions, and the duty cycle—the latter defined (different for scanners than for gears) as the ratio between the active (used effectively in the scanning process) and the total time.

Dynamics of the PM were also approached [21]. As their rotating speed ω goes up to 60 krpm, FEA (Finite Element Analysis) have to be performed to verify structural integrity issues and, most important, to determine deformations during the functioning regime at maximum speed (stress distribution for a 6 facets PM—Fig. 5b). Both facet-to-facet and facet-to-pivot deformations are important, as they may produce errors in the positioning of the beam. As a remark, precisions of $\lambda/8$ are required at manufacturing for the PM facets, where λ is the laser wavelength that will work with the scanner (in the range of 400–1400 nm, in general).

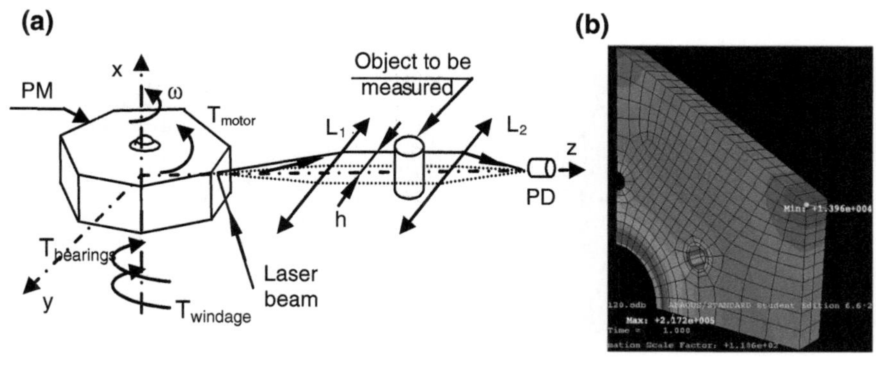

Fig. 5 PM scanner: **a** setup for on-line dimensional measurements; **b** example of FEA

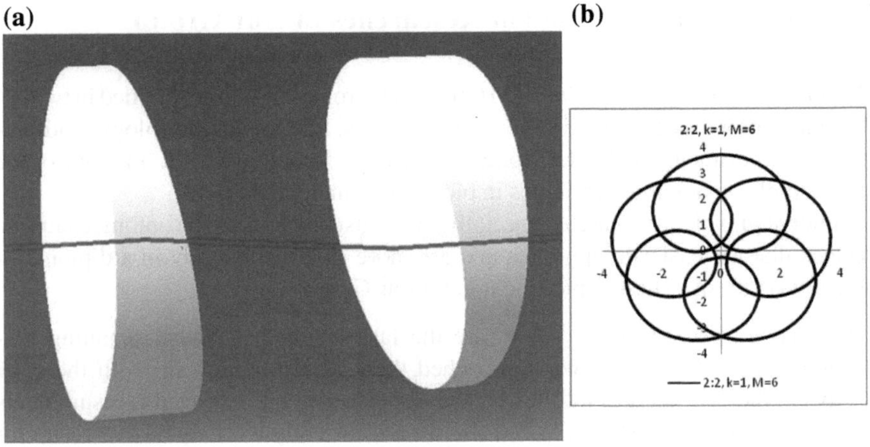

Fig. 6 **a** Modeling in CATIA of an optomechanical scanner with Risley prisms (optical wedges); **b** example of a scan pattern generated with two identical prisms (k $= \theta_2/\theta_1 =$ 1—ratio of the wedges angles) and M $= \omega_2/\omega_1 =$ 6—ratio of the rotational velocities (approach similar to the theory of mechanisms, but notations specific to the field of laser scanning) [14]

(ii) *Galvanometer-based scanners (GSs)*, the most used type of scanners [11], will be presented in [22]. They were introduced as an essential topic in Instrumentation (lecture and labs, 4th year, EE (Electrical Engineering) UG)—at UAVA.

(iii) *Risley prisms scanners* (Fig. 6) [11, 17] are approached mostly with UGs, while the previous PM and GS devices are usually topics for PG students. Strong conceptual similarities exist for Risley prisms with the classical Mechanism approach of gears, although the notations are different, as described in Fig. 6;

(iv) *Optomechanical choppers* (Fig. 7). These rather simple, very common devices in Photonics, that are utilized to chop a laser beam to generate series of impulses

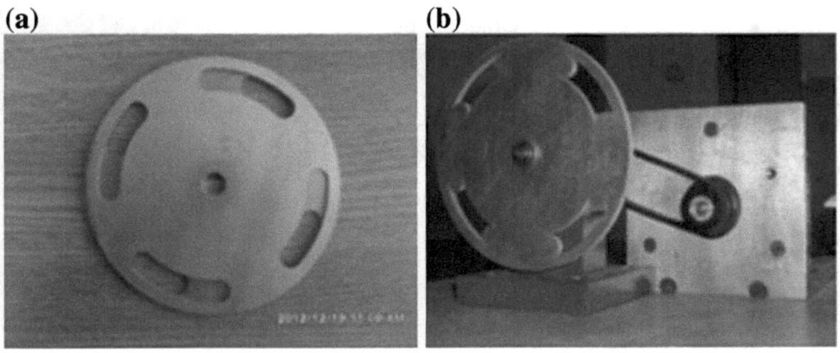

Fig. 7 Development of optomechatronic choppers [15] with different types of rotating wheels: **a** double wheel with windows with outward semi-circular edges; **b** assembly with a different type of wheel—with windows with inward edges (patent pending) [16]

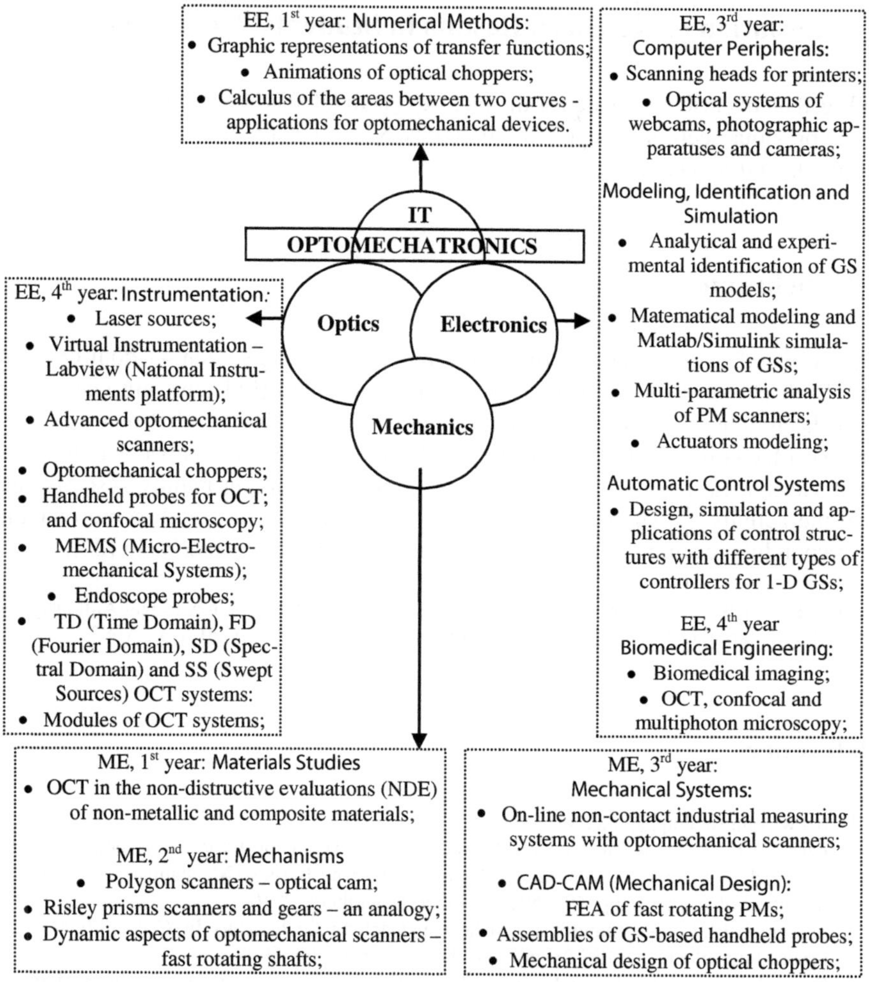

Fig. 8 Applications of the research topics in the UG curricula at UAVA—ME and EE

of adjustable frequency and profile were approached in our group first theoretically [15], for top-hat laser beams (the most used in laser manufacturing)—as there has been no unitary theory to embrace all their functioning cases. Prototype wheels (both classical and un-conventional, the latter with patent pending) and modules were then developed, constructed and tested [16], involving also—for the experimental part—student projects, some under way.

(v) *Optomechanical attenuators* were developed [23], with translational Risley prisms. Studies were carried out on the analysis and design of the optical part, and on the mechanical design, the latter both with conventional micrometer heads and with custom made piezo-electrical nano-translation stages.

4 Translation of Research Results in Students' Curricula

In Fig. 8 an overview is made for the lectures (with labs) where the topics presented in the previous section are applied—in italics. Several examples of the chapters and lab works introduced in the current UG curricula are pointed out—with bullets.

5 Conclusions

We presented a brief overview of our researches involving students—both UG and PG. Both classical mechanisms aspects were presented (linkages and gears) and inter-disciplinary approaches, that can be defined best as Optomechatronics, as they involve Mechatronics, but also Optics and Photonics. As this paper was focused on devices, a further presentation of these aspects, involving mainly systems and also the activity of our partners in current research projects—with some of their corresponding teaching for UG and training activity for PG—is done in [22].

Acknowledgments This work was supported by a grant of the Romanian National Authority for Scientific Research, CNDI–UEFISCDI project number PN-II-PT-PCCA-2011-3.2-1682.

References

1. www.3omgroup.appspot.com/
2. Duma, V.F.: Teaching Mechanisms: from Classical to Hands-on-Experiments and Research-Oriented. In: Pisla, D., Ceccarelli, M., Husty, M., Corves, B. (eds.) Mechanisms and Machine Science, vol. 5, Part 8, pp. 493–501. Springer, Heidelberg (2010)
3. Soylemez, E.: Using computer spreadsheets in teaching mechanisms. In: Ceccarelli, M. (ed.) Proceedings of EUCOMES 08, pp. 45–53. Springer, Berlin (2009)
4. Duma, V.F., Berar, T., Molnar, V., Patrusel, V.: Planar mechanisms developed in the mechanism laboratory of the Aurel Vlaicu University of Arad, In: Proceedings of ISREIE pp. 125–133 (2008)
5. Eckhardt, H.D.: Kinematic Design of Machines and Mechanisms. McGraw-Hill, New York (1998)
6. Perju, D.: Mecanisms of Fine Mechanics. Politehnica, Timisoara (1990)
7. Schitea, A., Duma, V.F.: Kinematic analysis of linkages using three different methods. In: Proceedings of 4th ISREIE, Series: Engineering (2012)
8. Duma, V.F., Patrusel V.: AutoCad simulations and experiments on the manufacturing of gears. In: Visa, I. (ed.) SYROM 2009: Proceedings of the 10th IFToMM International Symposium on Science of Mechanisms and Machines, pp. 529–538. Springer, Berlin (2010)
9. Berar, T., Duma, V.F.: Devices of orientation of pieces with irregular outline. Sci. Tech. Bull. Aurel Vlaicu Univ. Arad Mech. Eng. **3**, 28–40 (2007)
10. Duma, V.F. (ed.): Tools, devices and control apparatuses. Experimental works (in Romanian). Aurel Vlaicu University, Arad (2008)
11. Duma, V.F., Rolland, J.P., Podoleanu, A.Gh.: Perspectives of scanning in OCT. In: Proceedings of SPIE, vol. 7556, 7556–10, San Francisco (2010)

12. Oh, W.Y., Yun, S.H., Tearney, G.J., Bouma, B.E.: 115 kHz tuning repetition rate ultrahigh-speed wavelength-swept semiconductor laser. Opt. Lett. **30**, 3159–3161 (2005)
13. Yun, S.H., Boudoux, C., Tearney, G.J., Bouma, B.E.: High-speed wavelength-swept semiconductor laser with a polygon-scanner-based wavelength filter. Opt. Lett. **28**, 1981–1983 (2003)
14. Schitea, A., Tuef, M., Duma V.F.: Modeling of Risley prisms devices for exact scan patterns. In: Proceedings of SPIE, vol. 8789, 8789–40, Munchen (2013)
15. Duma, V.F.: Theoretical approach on optical choppers for top-hat light beam distributions. J. Opt. A Pure Appl. Opt. **10**, 064008 (2008)
16. Duma, V.F.: Prototypes and modulation functions of classical and novel configurations of optical chopper wheels. Lat. Am. J. Solids Struct. **10**(1), 5–18 (2013)
17. Marshall, G.F. (ed.): Handbook of Optical and Laser Scanning. Marcell Dekker, New York (2004)
18. Sweeney, M.N.: Polygon scanners revisited. In: Proceedings of SPIE, vol. 3131, pp. 65–76 (1997)
19. Walters, C.T.: Flat-field postobjective polygon scanner. Appl. Opt. **34**, 2220–2225 (1995)
20. Duma, V.F.: Novel approaches in the designing of the polygon scanners. In: Proceedings of SPIE, vol. 6785, p. 67851Q (2007)
21. Duma, V.F., Rolland, J.P.: Mechanical constraints and design considerations for polygon scanners. In: Pisla, D., Ceccarelli, M., Husty, M., Corves, B. (eds.) Mechanisms and Machine Science, Vol. 5, Part 8, pp. 475–483. Springer, Berlin (2010)
22. Duma, V.F., et al.: Multi-disciplinary optomechatronics student-included research with applications in biomedical imaging. In: Proceedings of ISEMMS (2013, in print)
23. Duma, V.F., Nicolov, M.: Neutral density filters with Risley prisms: analysis and design. Appl. Opt. **48**(14), 2678–2685 (2009)

Teaching Shaft Balancing as a Parameter Estimation Problem

J. Aginaga, X. Iriarte and J. Ros

Abstract One of the typical topics on Mechanisms and Machine Theory subject is shaft or rotor balancing. The unbalance is produced by an inadequate mass distribution of the rotating part. It can be said that static unbalance exists when the mass centre does not lie on the rotation axis and that dynamic unbalance happens when the rotating axis is not a principal inertia axis of the shaft. Making use of this point of view, the way to correct an unbalanced shaft is to cancel its inertia products involving the rotation axis. If these inertia products are known, it is easy to calculate the mass to be added or eliminated in order to cancel them. Instead, if the inertia products are unknown, they must be experimentally estimated in order to balance the shaft. This work presents a way to balance an unknown shaft as a parameter estimation problem. The formulation of least-square parameter estimation has been used in order to estimate the inertial properties of a shaft. Then, the balancing of the shaft is easily carried out. This approach has been explained to students of a Degree on Mechanical Engineering at the Public University of Navarra. By means of a virtual model of an unbalanced shaft, students have estimated the inertial properties of the unbalanced shaft and then calculated the masses to be added in order to balance it. I has been a satisfactory experience in order to improve the students skills on the Mechanism and Machine Theory and they have also learned a formulation that can be used for other kind of problems.

Keywords Shaft balancing · Teaching · Inertia loads · Unbalanced rotor · Parameter estimation

J. Aginaga(✉) · X. Iriarte · J. Ros
Public University of Navarra, Navarra, Spain
e-mail: jokin.aginaga@unavarra.es

X. Iriarte
e-mail: xabier.iriarte@unavarra.es

J. Ros
e-mail: jros@unavarra.es

J. C. García-Prada and C. Castejón (eds.), *New Trends in Educational Activity in the Field of Mechanism and Machine Theory*, Mechanisms and Machine Science 19,
DOI: 10.1007/978-3-319-01836-2_13,

1 Introduction

Shaft balancing is a common topic in Mechanisms and Machine Theory subject. It consists of correcting the inertial properties of an unbalanced shaft in order to reduce or eliminate undesired inertia forces and moments. These unwanted loads are produced by static or dynamic unbalance. The former occurs when the mass centre of the shaft does not stand in the rotation axis. Regarding to the latter, it is customary to represent it as a shaft with two equal masses placed at opposite ends of the shaft and at equal distances from the rotation axis. The inertia forces of the masses form a couple which rotates with the masses. This couple causes a moment on the ground plane, alternately lifting and dropping the left and right ends of the shaft. This is the way static and dynamic balance is explained in the most of classical books of Mechanisms and Machine Theory [1–3].

However, from another point of view, adding these two masses changes the principal inertia axes and consequently the rotation axis does not lie on a principal inertia axis. Then, it can be said that dynamic unbalance is produced when inertia products involving the rotating axis are not null [4].

A usual way to balance a shaft is adding two mases in two different known planes. The masses and their location are calculated in order to eliminate said reaction forces and moments, that is, in order to move the mass centre to the rotation axis and to cancel inertia products involving the rotation axis.

When the characteristics of the shaft to be balanced are known, location of its mass centre and inertia products are known and it is easy to calculate the balancing masses. By contrast, when the inertia characteristics of the unbalanced shaft are unknown, they must be calculated by means of experimental procedures. There are several methods and balancing machines which calculate the balancing masses to be added. Based on least-squares parameter estimation techniques [5], a new approach to calculate the inertia characteristics of an unbalanced shaft has been developed. The use of parameter estimation techniques leads to an estimation of inertia products, location of mass centre and even the mass itself, based on the measurements of reaction forces at joints when the shaft rotates at constant velocity. Such reaction forces must be measured in at least two instants of the experiment and the more measurements are taken the more accurate the calculation is.

This new procedure of calculating balancing masses has been explained to students of a Degree on Mechanical Engineering at the Public University of Navarra. It has been a very positive experience and students have had the possibility to deepen their knowledge in mechanics and to review the underlying concepts.

2 Static and Dynamic Balancing

The motion of a mechanism usually involves dynamic loads at joints and supports due to the dynamic behaviour of the mechanism parts. These forces can cause vibrations having a negative influence over the parts integrity, leading to wear at joints and

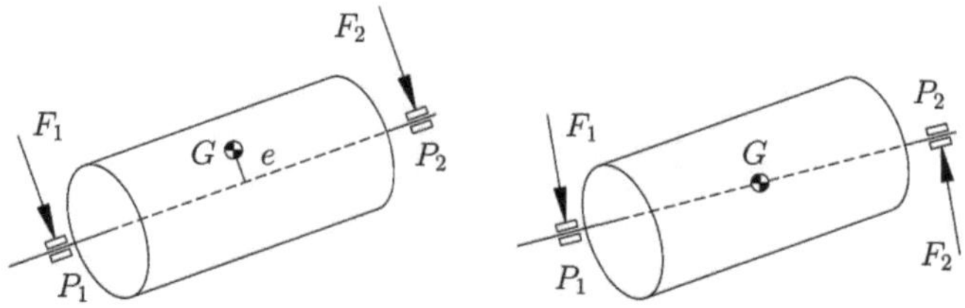

Fig. 1 Static and dynamic of unbalance

even fatigue failure of the parts. Balancing consist of correcting or eliminating these undesired loads.

Any element in pure rotation can theoretically be perfectly balanced to eliminate these forces and moments. Then, all the rotating elements of a machine are usually balanced. Balancing involves designing elements or sets of elements, with an adequate distribution of mass so that unbalance reactions are avoided. Although balancing is taken into account when designing a rotating part, it is often necessary to perform an experimental balancing in order to correct manufacturing errors. Moreover, after some cycles of work, a new balancing is also usual in order to correct wear or temperature effects.

There are two types of unbalance: static and dynamic unbalance. Static unbalance occurs when the centre of inertia of the rotor does not lie on its axis of rotation. In dynamic unbalance, the rotation axis does not lie on a principal inertia axis. Hence, it can be said that dynamic unbalance is produced when inertia products involving the rotating axis are not null. Figure 1 shows illustrative examples of both static and dynamic unbalance.

As shown in Fig. 1, static unbalance produces similar reactions on both supports while dynamic unbalance produces opposite reactions.

The way to correct the static or dynamic unbalances is by adding or eliminating mass. Regarding the static unbalance, the objective is to move the inertia centre to the rotating axis. Meanwhile, the dynamic balancing aims to align a principal axis of inertia with the rotating axis and this is made by cancelling the inertia products involving the rotating axis.

A common shaft will have both static and dynamic unbalance. Then, mass must be added or eliminated in such a way that both unbalancing types are corrected. A usual approach is to add two point masses in two predefined correction planes of the rotating shaft. The planes are usually massless discs with a defined radius and the parameters to be calculated are the masses themselves and an angular coordinate which determines their location. Figure 2 shows a shaft with two green correction planes.

The masses to be added can be called m_L and m_R and their location at a known radius is defined by angles ϕ_L and ϕ_R. Figure 3 shows front and lateral views of the unbalanced shaft with balancing masses located at generic ϕ_L and ϕ_R angles.

If the shaft inertial parameters are known, that is, its mass centre location and inertia products, balancing masses can be easily calculated.

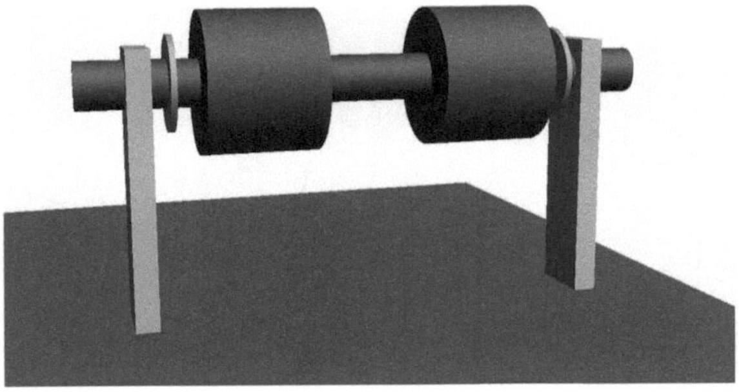

Fig. 2 A shaft with balancing planes

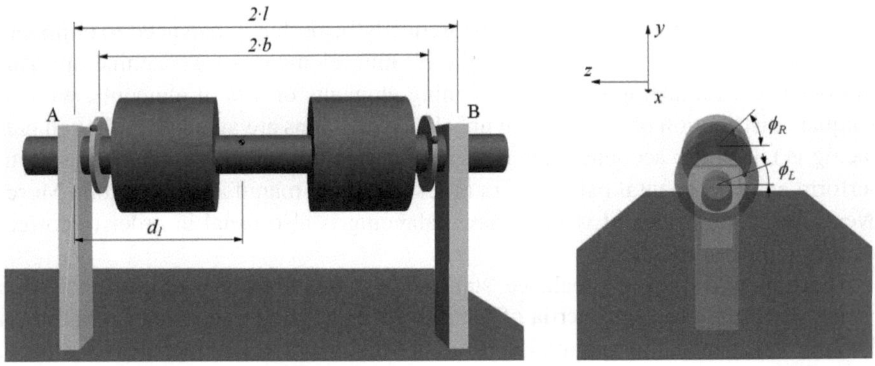

Fig. 3 Location of balancing masses

First, the mass centre of the new shaft (i.e., the unbalanced shaft with balancing masses) must be located in the rotating axis. If the rotating axis is denoted as x axis, Eq. 1 constraints the location of the mass centre of the balanced shaft.

$$\begin{aligned} m \cdot y_G + m_L \cdot r \cdot sin\phi_L + m_R \cdot r \cdot sin\phi_R &= 0 \\ m \cdot z_G - m_L \cdot r \cdot cos\phi_L - m_R \cdot r \cdot cos\phi_R &= 0 \end{aligned} \tag{1}$$

where m is the mass of the unbalanced shaft, y_G and z_G are the coordinates of its mass centre in a frame fixed to the shaft and r is the radius of the balancing discs.

The balancing is completed by constraining inertia products involving the rotating axis to be equal to zero, as shown in Eq. 2

$$\begin{aligned} I_{12} - m_L \cdot (l - b) \cdot r \cdot sin\phi_L - m_R \cdot (l + b) \cdot r \cdot sin\phi_R &= 0 \\ I_{13} + m_L \cdot (l - b) \cdot r \cdot cos\phi_L + m_R \cdot (l + b) \cdot r \cdot cos\phi_R &= 0 \end{aligned} \tag{2}$$

where I_{12} and I_{13} are the inertia products of the unbalanced shaft with respect to point A and l and b are the half of the distance between supports and the half of the distance between balancing planes respectively (Fig. 3).

By soving Eqs. 1 and 2, values for balancing masses m_L and m_R and angles ϕ_L and ϕ_R of their location are obtained.

3 Balancing a Shaft with Unknown Inertial Properties

Usually the inertial properties of the shaft are unknown due to manufacture errors, deformations produced when transporting them or high operating temperatures. Then, the mass centre location and the inertia products must be calculated by means of experimental procedures. We choose the reaction forces at supports to be measured during the operation of the shaft. If the rotation axis is denoted as x, reactions at supports are measured in y and z directions.

In order to calculate the inertial parameters (mass, mass centre location and inertia tensor) of the unknown shaft, dynamics of the rotating shaft are set out. From the equilibrium of forces Eq. 3 is obtained:

$$\begin{Bmatrix} 0 \\ f_y^A \\ f_z^A \end{Bmatrix}_{xyz} + \begin{Bmatrix} 0 \\ f_y^B \\ f_z^B \end{Bmatrix}_{xyz} + \begin{Bmatrix} 0 \\ -m \cdot g \\ 0 \end{Bmatrix}_{xyz} + \begin{Bmatrix} 0 \\ m \cdot d_2 \cdot \omega^2 cos\theta - m \cdot d_3 \cdot \omega^2 sin\theta \\ m \cdot d_2 \cdot \omega^2 sin\theta + m \cdot d_3 \cdot \omega^2 cos\theta \end{Bmatrix}_{xyz} = \begin{Bmatrix} 0 \\ 0 \\ 0 \end{Bmatrix}_{xyz} \tag{3}$$

where f_x^A, f_z^A, f_x^B and f_z^B are the reactions measured by the sensors, θ is the angular coordinate of the rotation of the shaft, ω is the angular velocity in which measurements are taken ($\omega = \dot{\theta}$), and d_2 and d_3 define the coordinates of the mass centre of the unknown shaft in y and z axes, respectively, with respect to point A.

Equilibrium of moments with respect to point A leads to Eq. 4.

$$\begin{Bmatrix} 0 \\ -2 \cdot l \cdot f_z^B \\ 2 \cdot l \cdot f_y^B \end{Bmatrix}_{xyz} + \begin{Bmatrix} -m \cdot g \cdot d_2 sin\theta + m \cdot g \cdot d_3 cos\theta \\ 0 \\ -m \cdot g \cdot d_1 \end{Bmatrix}_{xyz} + \begin{Bmatrix} M \\ 0 \\ 0 \end{Bmatrix}_{xyz} + \begin{Bmatrix} 0 \\ I_{13} \cdot \omega^2 \cdot cos\theta + I_{12} \cdot \omega^2 \cdot sin\theta \\ I_{13} \cdot \omega^2 \cdot sin\theta - I_{12} \cdot \omega^2 \cdot cos\theta \end{Bmatrix}_{xyz} = \begin{Bmatrix} 0 \\ 0 \\ 0 \end{Bmatrix}_{xyz} \tag{4}$$

Equations 3 and 4 present a system of equations representing the dynamics of the rotating shaft. It is worthy to note that the first equation of Eq. 3 is null. Also, Eq. 4 introduces the actuation moment M, which is unknown but unnecessary for the objective of calculating the inertial parameters of the shaft. Consequently, the first equation of Eq. 4 is not taken into account. Then, the remaining system is composed of four equation and six unknowns, namely, m, d_1, d_2, d_3, I_{xy} and I_{xz}. These six unknowns are the inertial parameters of the unknown shaft.

Since the equation system is not a linear system with respect to some of the unknowns, another set of unknowns is chosen in order to make it linear. The new set of unknowns is: $m, md_1 = m \cdot d_1, md_2 = m \cdot d_2, md_3 = m \cdot d_3, I_{12}$ and I_{23}. With these new set of unknowns, the system can be rewritten as in Eq. 5.

$$\begin{pmatrix} -g & 0 & \omega^2 cos\theta & -\omega^2 sin\theta & 0 & 0 \\ 0 & 0 & \omega^2 sin\theta & \omega^2 cos\theta & 0 & 0 \\ 0 & 0 & 0 & 0 & -\omega^2 sin\theta & \omega^2 cos\theta \\ 0 & -g & 0 & 0 & \omega^2 cos\theta & -\omega^2 sin\theta 0 \end{pmatrix} \begin{Bmatrix} m \\ md_1 \\ md_2 \\ md_3 \\ I_{12} \\ I_{23} \end{Bmatrix} = \begin{Bmatrix} -f_y^A - f_y^B \\ -f_z^A - f_z^B \\ 2 \cdot l \cdot f_z^B \\ -2 \cdot l \cdot f_y^B \end{Bmatrix} \tag{5}$$

The system of Eq. 5 is linear so it can be rewritten by means of a matrix $\mathbf{A}_i$, a vector of unknowns $\mathbf{p}$ and a vector of known terms $\mathbf{b}_i$. If measurements are taken in the i-th instant, equation can be written as:

$$\mathbf{A}_i \cdot \mathbf{p} = \mathbf{b}_i \tag{6}$$

Since the equation system of Eq. 6 has four equations and six unknowns, it is necessary to rewrite it for at least two instants not to have an underdetermined system of equations. Setting out the system of Eq. 6 for several instants, an overdetermined linear system of equations is obtained. Using least-square techniques, an estimation of the parameters is calculated. Obviously, the more instants are considered the more accurate solution will be Eq. 7 shows the system to be solved for n instants, each of them corresponding to a certain value of coordinate θ.

$$\begin{pmatrix} \mathbf{A}_1 \\ \vdots \\ \mathbf{A}_n \end{pmatrix} \mathbf{p} = \begin{pmatrix} \mathbf{b}_1 \\ \vdots \\ \mathbf{b}_n \end{pmatrix} \Rightarrow \mathbf{A}\,\mathbf{p} = \mathbf{b} \tag{7}$$

Solving Eq. 7 with the pseudo-inverse of $\mathbf{A}$, values of the inertial parameters of the unknown shaft are obtained. Once these are known, balancing masses (m_L and m_R) and their angular location (ϕ_L and ϕ_L) in the balancing planes are calculated by means of Eqs. 1 and 2 so that the shaft is balanced.

4 Practical Exercise

As shown in Sect. 3, the inertial properties of a shaft can be estimated by measuring reaction forces in the supports and using the dynamic equations of the shaft. Since it is not possible to have real shafts operating at the laboratories of the university, a model of an unbalanced shaft is developed in 3D_Mec [6, 7]. Figure 4 shows a picture of the unbalanced shaft of the 3D_Mec model. Two red spheres are depicted representing the unbalance of the shaft and two red arrows represent reaction forces at supports at certain instant.

The 3D_Mec model is given to the students and, by executing it, virtual measurements of reaction forces are exported into a MatLab file. Students have to apply parameter estimation formulation in order to program the equation system of Eq. 7. By soving it, they obtain the inertial parameters of the shaft.

Once the inertial parameters of the shaft have been estimated, the balancing masses must be calculated. For that purpose, students must program Eqs. 1 and 2 constraining the mass centre to be in the rotation axis and inertia products involving the rotation axis to be null.

The last step is to check the balancing masses in the 3D_Mec model. When running the simulation including the balancing masses, the reaction forces in the bearings should be reduced to gravity forces, so that the red arrows representing supports reactions are vertical and constant.

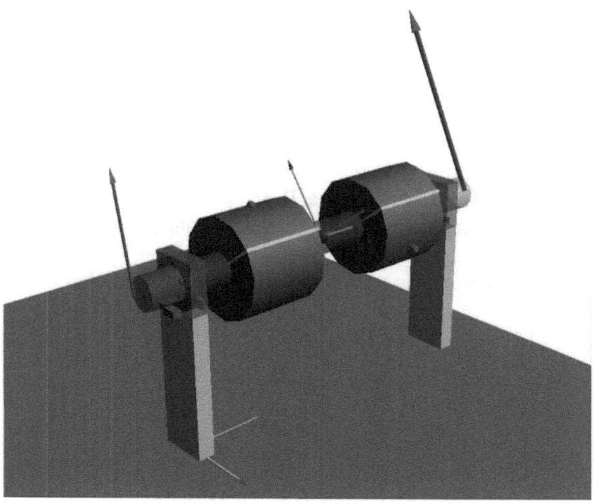

Fig. 4 3D_Mec model of a unbalanced shaft

5 Conclusions

A new approach to balance unbalanced shafts has been presented. The dynamic unbalance is shown to be due to a misalignment of the rotation axis with respect to a principal inertia axis. Then, the way to align them is to cancel the inertia products involving rotation axis by adding two masses in certain predefined planes, so that the principal axis gets aligned with the actual rotation axis. When the inertial properties of the shaft to be balanced are unknown, these are calculated by means of least-squares parameter estimation formulation.

This approach to understand unbalancing has been explained to engineering students in Mechanism and Machine Theory subject. The fact that they have already dealt with inertia tensors and principal inertia axes in Mechanics subject has made easier for them to understand the basis of unbalanced shafts and the way they are balanced.

By means of a 3D_Mec model of an unbalanced shaft, students have managed their skills with different softwares establishing and using their knowledge in shaft balancing. Furthermore, students have learned the least-squares parameter estimation formulation which is actually used to solve problems in other fields, such as mechanical model validation in robotics, vehicle dynamics, etc.

References

1. Shigley, J.E., Uicker, J.J. Jr.: Theory of machines and mechanisms. McGraw-Hill (1981)
2. Norton, R.L.: Design of machinery. McGraw-Hill (1992)
3. Sandor, G.N., Erdman, A.G.: Advanced mechanism design. Prentice-Hall, Inc (1984)
4. Cardona Foix, S., Clos Costa, D.: Teoría de máquinas. Edicions UPC (2001)
5. van den Bos, A.: Parameter estimation for scientists and engineers, Wiley (2007)
6. Ros, J., Gil, J. and Zabalza, I.: 3D_Mec: An application to teach mechanics. In: ASME International Design Engineering Technical Conferences, IDETC'05. California, USA, pp. 1895–1899 (2005)
7. http://www.imac.unavarra.es/3d_mec/pages/en.php

Paradox in the Mechanism Science

K. S. Ivanov and K. Jilisbaeva

Abstract Paradox of the mechanism theory has been detected in connection with attempts of practical use mechanism with two degree of freedom and one entry for achievement of power adaptation. According to the mechanism theory and machines such mechanism is statically indefinable. However recently there were patents and publications with the description of efficient two mobile adaptive mechanisms with one entry. Paradox of the mechanism theory consists in the following: the kinematic chain with two degree of freedom and one entry containing the closed contour is statically and kinematically definable. In Ivanov's works the proof of definability of two mobile kinematic chain motion is resulted on the basis of additional analytical constraint by a virtual works principle. A certain geometrical image comparable to design there should match to analytical additional constraint. The present work is devoted to disclosing of a geometric constraint of parameters of kinematic chain with two degree of freedom and to statement of the mechanism theory paradox in geometrical and an analytic form. Work is executed on the basis of regularities of mechanics and mechanism theory.

Keywords Mechanism science · Paradox · Force adaptation

1 Introduction

Paradox of the mechanism theory has been detected in connection with attempts of practical use mechanism with two degree of freedom and one entry for achievement of self-regulation or power adaptation. According to the mechanism theory [1] such

K.S. Ivanov (✉)
Almaty University of Power Engineering and Telecommunication, Almaty, Kazakhstan
e-mail: ivanovgreek@mail.ru

K. Jilisbaeva
Kazakh National University, Almaty, Kazakhstan

J. C. García-Prada and C. Castejón (eds.), *New Trends in Educational Activity in the Field of Mechanism and Machine Theory*, Mechanisms and Machine Science 19,
DOI: 10.1007/978-3-319-01836-2_14, © Springer International Publishing Switzerland 2014

mechanism is statically indefinable as the degree of freedom number should be equal to input links number. However recently there were patents and publications with the description of efficient two mobile self-controlled mechanisms with one entry. Paradox of the mechanism theory consists in the following: the kinematic chain with two degree of freedom and one entry containing the closed contour is statically and kinematic definable.

In patents of Croquet [2] and Volkov [3] the idea of creation stepless adjustable transfer with constant cogging of toothed wheels in the form of the hydrodynamic converter and gear differential with two degree of freedom is presented. The mechanism with two degree of freedom and one entry in the form of gear differential is presented in Harris's patent [4]. Ivanov's patents [5, 6] provides the use of inertia properties at start-up and contains regularities of differential mechanism synthesis on the set range of transfer ratios.

Existing practical workings out in the form of patents for inventions are intuitive as a rule. They do not contain the theoretical proof of definability of two mobile mechanisms containing only one entry.

Ivanov has resulted the proof of definability of two mobile mechanism motion on the basis of virtual works principle [7]. Definability of motion of two mobile mechanisms with one entry leads to occurrence of basic new phenomenon—effect of power adaptation. However use of virtual works principle should be proved—the mechanism should have a certain additional constraint between links.

In works [8, 9] additional constraint between links of two mobile mechanism with one entry has been theoretically proved. Analytical additional constraint represents a certain abstraction. Basic new aspect of constraint should have the certain geometrical image comparable to a design. The present work is devoted to disclosing of a geometric constraint of parameters of the kinematic chain with two degree of freedom and to statement of essence of paradox of the mechanism theory in geometrical and an analytic form. Work is executed on the basis of regularities of mechanics and mechanism theory.

2 Research of Motion of the Kinematic Chain with Two Degree of Freedom

The kinematic chain with two degree of freedom is presented in the form of a differential wheelwork with two carriers (Fig. 1).

The mechanism contains rack 0, input carrier H_1, input satellite 2, block of the central wheels 1-4, block ring wheels 3-6, the output satellite 5 and output carrier H_2. Toothed wheels form four link closed contour 1-2-3-6-5-4. Sizes of toothed wheels 1, 2, 3, 4, 5, 6 are defined by matching radiuses $r_i \quad i = 1, 2, 3, 4, 5, 6$. Radiuses of carriers $r_{H1} = r_1 + r_2$, $r_{H2} = r_4 + r_5$.

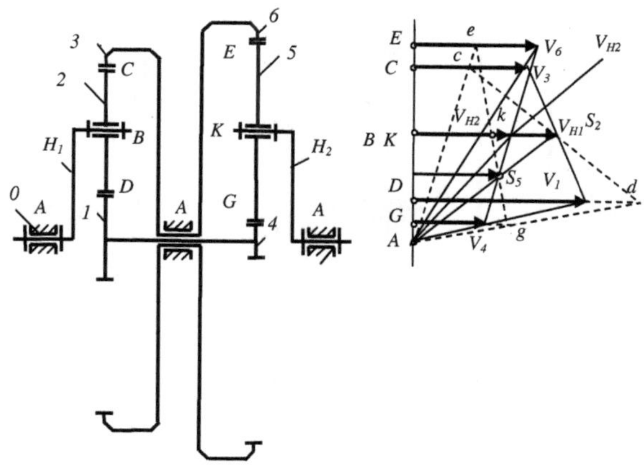

Fig. 1 Gear differential mechanism and a picture of its speeds

To the right of mechanism the picture of speeds V_i $i = 1, 2, 3, 4, 5, 6$ of mechanism links is presented. Linear speeds v_i are expressed through angular speed ω_i by formula $V_i = \omega_i r_i$. Linear speeds of carriers $V_{Hi} = \omega_{Hi} r_{Hi}$, $i = 1,\ 2$.

The initial picture of speeds is shown by full lines. Intermediate picture of speeds matches to change of initial parameters of motion at invariable input speed and is shown by dashed lines.

On Fig. 1 it is visible that instantaneous center of turn S_5 of satellite 5 occupies invariable position on a line Ab of input carrier angular speed.

Let's prove that at constant angular speed of input carrier the S_5 of satellite 5 will occupy invariable position on a line Ab.

Theorem: In the differential wheelwork containing two carriers, two satellites and two blocks of the central wheels the satellite has the constant instantaneous center of relative turn on the opposite carrier.

For the proof we will observe at first picture of speeds of the kinematic chain in inverse motion at motionless carrier H_1 (Fig. 2).

At motionless carrier H_1 linear speed $V_{H1} = 0$. The instantaneous center of turn S_2 of satellite 2 coincides with point B. We will run a line cS_2d angular speed of satellite 2. We will run a line Ac angular speed of a link 3-6 and it is found on it a point e defining speed V_6 of mechanism point E. We will run a line Ad angular speed of link 1-4 and it is found on it a point g defining speed V_4 of mechanism point G. We will run a line eg angular speed of satellite 5.

Intersection point S_5 of a line eg with a vertical line of zero speeds is the instantaneous center of turn of satellite 5. Point S_5 is simultaneously instant centre of turn of satellite 5 concerning motionless carrier H_1. The position of point S_5 (size $y = BS_5$) can be defined through mechanism geometrics.

$$y = r_5 \frac{r_1 r_6 - r_3 r_4}{r_1 r_6 + r_3 r_4}. \tag{1}$$

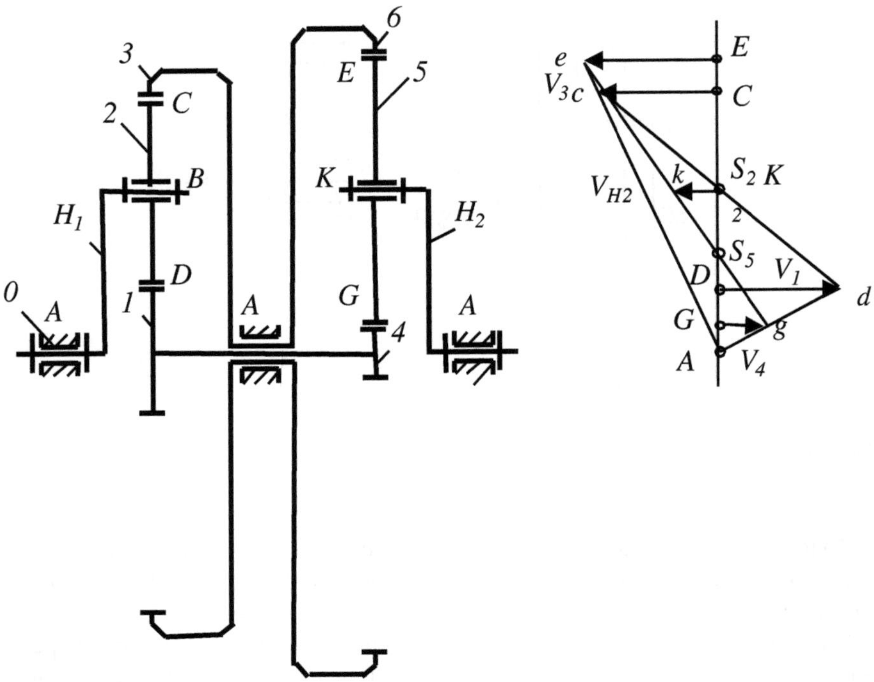

Fig. 2 Gear differential mechanism and picture of it speeds in inverse motion

Equation (1) one-valued defines position of satellite 5 instantaneous center of turn S_5 in inverse motion concerning the motionless carrier H_1. In mechanism valid motion the point S_5 will move together with the carrier H_1 as it is shown in picture of speeds of moving mechanism (Fig. 1).

Hence the satellite 5 has the instantaneous center of relative turn on the opposite carrier H_1 as was to be shown.

The invariable position of the instantaneous center of relative turn S_5 on carrier H_1 defines an additional geometric constraint which occurs in the kinematic chain with two degree of freedom. However this additional constraint has no constructive performance. Mechanically the mechanism remains the kinematic chain with two degree of freedom and keeps the properties.

Equation (1) expresses analytically this additional geometric constraint. As a result the kinematic chain becomes definable at presence only one input link. It means that at the set angular speed of input carrier it is possible to one-valued define speeds of all other links and mechanism points.

The additional geometric constraint leads to transformation of the initial kinematic chain to the conditional mechanism with one degree of freedom. The found regularity allows building replacing mechanism (Fig. 3) with the changed structure. However this constructive additional constraint is passive. Actually the degree of freedom number remains equal to two.

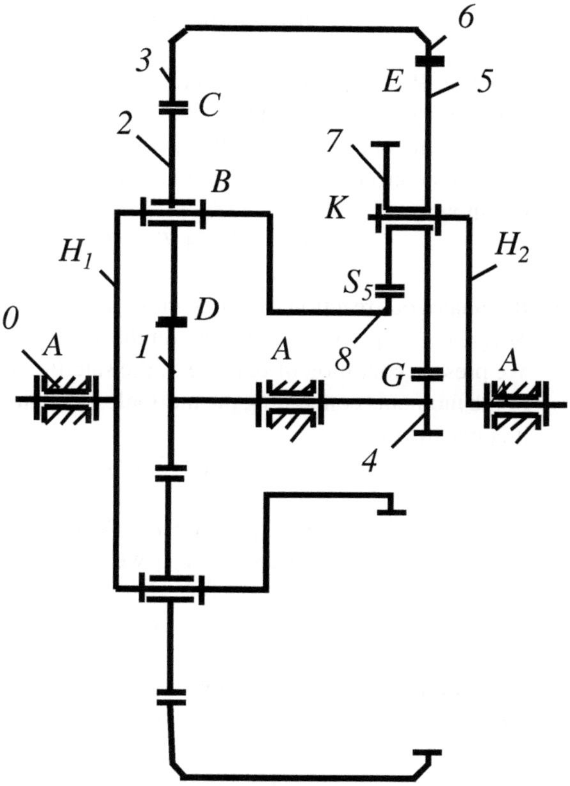

Fig. 3 The conditional replacing mechanism with one degree of freedom

3 Power Analysis of the Kinematic Chain with Two Degree of Freedom

In the presence of passive structural constraint and possibility of internal mobility of links of the closed contour according to picture of speeds (Fig. 2) it is necessary to recognize, first, possibility of definability of motion and, secondly, to install the reason of possible internal mobility of links of a contour.

The considered kinematic chain with two degree of freedom contains two links, having generalized co-ordinates (carriers H_1 and H_2) and the closed contour affiliated to them containing toothed wheels 1-2-3-6-5-4 and having zero mobility.

Statement of the power analysis problem.

Two generalized forces M_{H1}, M_{H2} (moments of forces on carriers H_1 and H_2) are given.

It is necessary to define reactions in kinematic pairs.

The solution.

(1) From carriers H_1 and H_2 on satellites 2 and 5 are transferred the reactions

$$R_{H1} = M_{H1}/r_{H1}, \quad R_{H2} = M_{H2}/r_{H2}. \tag{2}$$

(2) Reaction R_{H1} in point B acts and reactions $R_{12} = R_{32} = R_{H1}/2 = M_{H1}/2r_{H1}$ in points D and C acts on the satellite 2.
We present condition of satellite 2 balance in the form of equality to null of the moments concerning the motionless instant centre of speeds S_{20} (on Fig. 2 it is not shown).

$$R_{12} \cdot DS_{20} + R_{32} \cdot CS_{20} = R_{H1} \cdot BS_{20}. \tag{3}$$

(3) It is analogous reaction R_{H2} in point K acts and reactions $R_{45} = R_{65} = R_{H2}/2 = M_{H2}/2r_{H2}$ in points G and E act on the satellite 5.
We present condition of balance of the satellite 5 in the form of equality to null of the moments concerning the motionless instant centre of speeds S_{50} (on Fig. 2 it is not shown).

$$R_{45} \cdot GS_{50} + R_{65} \cdot ES_{50} = R_{H2} \cdot KS_{50}. \tag{4}$$

(4) We will multiply the Eq. (3) on ω_2. With the account $\omega_2 \cdot DS_{20} = V_1$, $\omega_2 \cdot CS_{20} = V_3$, $\omega_2 \cdot BS_{20} = V_{H1}$ we will gain

$$R_{12} \cdot V_1 + R_{32} \cdot V_3 = R_{H1} \cdot V_{H1}. \tag{5}$$

Let's multiply the Eq. (4) on ω_5. With the account $\omega_5 \cdot GS_{50} = V_4$, $\omega_5 \cdot ES_{50} = V_6$, $\omega_5 \cdot KS_{50} = V_{H2}$ we will gain

$$R_{45} \cdot V_4 + R_{65} \cdot V_6 = R_{H2} \cdot V_{H2}. \tag{6}$$

Let's combine the Eq. (5) and (6)

$$R_{12} \cdot V_1 + R_{32} \cdot V_3 + R_{45} \cdot V_4 + R_{65} \cdot V_6 = R_{H1} \cdot V_{H1} + R_{H2} \cdot V_{H2}. \tag{7}$$

The Eq. (7) contains the forces acting on the closed contour containing toothed wheels 1-2-3-6-5-4 and connected to two carriers (generalized co-ordinates). The left side of Eq. (7) represents the sum of powers (or works) internal forces of the closed contour. The right side of Eq. (7) represents the sum of powers (or works) external forces of the closed contour. We will consider kinematic pairs of the closed contour as ideal. Then the sum of works (or powers) internal forces of the closed contour is equal to null

$$R_{12} \cdot V_1 + R_{32} \cdot V_3 + R_{45} \cdot V_4 + R_{65} \cdot V_6 = 0. \tag{8}$$

From the Eq. (8) follows that the sum of works of external forces of the closed contour also is equal to null

$$R_{H1} \cdot V_{H1} + R_{H2} \cdot V_{H2} = 0. \tag{9}$$

Taking into account Eq. (2) we will gain

$$M_{H1} \cdot \omega_{H1} + M_{H2} \cdot \omega_{H2} = 0. \tag{10}$$

That is the sum of powers (or works) generalized forces is equal to null. It means that all considered kinematic chain is in equilibrium.

From Eq. (10) follows that one of the generalized forces should be negative. Than it is necessary to consider, for example, that M_{H2} is the moment of resistance on output carrier H_2 and M_{H1} is the driving moment on the input carrier H_1.

Thus it will appear that in the kinematic chain with two degree of freedom and with the closed contour only one link (input carrier H_1) can have generalized co-ordinate (independent angular speed ω_{H1}). Such kinematic chain will have the kinematic definability as Eq. (9) allows to define the output angular speed ω_{H2} at the set moments M_{H2} and M_{H1} .

Hence the kinematic chain with two degree of freedom and with the closed contour at presence only one entry will have static and kinematic definability and will be the mechanism.

The mechanism with two degree of freedom possesses the surprising property outlawing from the Eq. (10)

$$\omega_{H2} = \frac{M_{H1} \cdot \omega_{H1}}{M_{H2}}. \tag{11}$$

Equation (11) expresses effect of power adaptation: at the set constant parameters of input power M_{H1} and input ω_{H1} the output angular speed ω_{H2} inversely proportional to the set variable output moment of resistance M_{H2}. The mechanism creating effect of power adaptation was named as the adaptive mechanism.

Work of the adaptive mechanism begins with start at which the mechanism moves as a single whole with one degree of freedom (thanks to presence of a friction in the absence of an external resistance). After start the external moment of resistance translates the mechanism in a condition with two degree of freedom with internal relative motion of links in the presence of effect of power adaptation.

Presence of effect of power adaptation defines a discovery in the field of mechanics.

The discovery formula: Effect of power adaptation in the mechanic field consists that the kinematic chain with two degree of freedom containing input link, output link and the closed mobile contour placed between them provides the motion of output link with a speed inversely proportional to loading on it at constant input power.

4 Conclusion

Paradox of the mechanism theory is solved. It is proved that the mobile closed mechanical contour of the kinematic chain with two degree of freedom creates additional constraint and provides static and kinematic definability of chain at presence only one entry. Such chain gets property of force adaptation.

The developed discovery allows providing the variable transfer relation only at the expense of use of the closed contour and its properties without application of control means. Mechanical properties of the closed contour allow to provide the demanded transfer ratio independently, stepless and automatically.

References

1. Levitsky, N.I.: Theory of mechanisms and machines. Mthe Science Russia (1979)
2. Crockett, S.J.: Shiftless, continuously-aligning transmission. Patent of USA 4,932,928, Cl. F16H 47/08, U.S. Cl. 475/51; 475/47, USA (1990)
3. Volkov, I.V.: Way of automatic and continuous change of a twisting moment and speed of twirl of the output shaft depending on a tractive resistance and the device for its realization. The invention description to the patent of Russia RU 2 234626, Russia (2004)
4. John, H.: Power transmission system comprising two sets of epicycle gears. Patent of Great Britain GB2238090 (A), UK (1991)
5. Ivanov, K.S., Yaroslavtseva, E.K.: Way of automatic and continuous change of a twisting moment and speed of twirl of the output shaft depending on a tractive resistance and the device for its realization. The patent of Russia RU 1 2398989. Russia (2010)
6. Ivanov, K.S.: Almaty, KAZ—Owner of the registered sample. The name—Device of automatic and continuous change of a twisting moment—and changes of a corrected speed of output shaft depending on a tractive resistance. The deed on registration of the registered sample 20 2012 101 273.1. Day of Registration 02.05. (2012). German patent and firm establishment. Federal Republic Germany (2012)
7. Ivanov, K.S.: The simplest automatic transfer box. WCE (2010). World Congress on Engineering 2010 (ICME). London, UK, pp. 1179–1184 (2010)
8. Ivanov, K.S.: Theorem about balance of the closed contour. The Mechanism theory and machines. Periodic Scientifically-Methodical J. **2**(16) (2010). The St.-Petersburg state polytechnical university, vol. 8, pp. 85–89
9. Ivanov, K.S.: Paradox of mechanics—a basis of creation CVT. Transactions of 2-d IFToMM Asian Conference on Mechanisms and Machines Science. Tokyo, Japan, pp. 245–264, 7–10 Nov 2012

Modern View on "History of Science" Teaching at Technical Universities

Olga Egorova, Alexander Evgrafov and Valeria Rayushkina

Abstract One of priority missions of "History of Science" teaching at technical universities is the promotion of scientists-engineers achievements and industrial heritage preservation of mankind. At the same time, the "History of Science" course in the students' program of training leads to expansion of their outlook and deep understanding of contiguity of nature development, engineering and society, as well as to increase of overall level of technical universities graduates.

Keywords History of science · Technical university · Historiography · SPSPU · BMSTU

1 A Brief "History of Science"

Humankind has always been inquisitive, needing to understand why things behave in a certain way, and trying to link observation with prediction. The development of any science as a way of studying and understanding the world, from the primitive stage of noting important regularities in nature to the epochal revolution in our notion of what constitutes reality that has occurred is reflected by the History of Science (HS). As all areas of knowledge it has its own history. Thus, for the first ever time a HS chair was created at the Collège de France in 1892 and Pierre Laffite became its first chair-holder.

O. Egorova (✉)
Bauman Moscow State Technical University (BMSTU), Moscow, Russia
e-mail: cuba2006@inbox.ru

A. Evgrafovand · V. Rayushkina
St. Petersburg State Polytechnical University (SPbSPU), St. Petersburg, Russia
e-mail: a.evgrafov@spbstu.ru

V. Rayushkina
e-mail: lerik-raj@mail.ru

J. C. García-Prada and C. Castejón (eds.), *New Trends in Educational Activity in the Field of Mechanism and Machine Theory*, Mechanisms and Machine Science 19,
DOI: 10.1007/978-3-319-01836-2_15,

In Russia the "History of Science" course started at Lomonosov Moscow State University (MSU) in 1903 with lectures of well-known historian of science V.I. Vernadsky (1863–1945). According to his words the beginnings of modern science which have brought later to the great rise of the seventeenth century, are equally based on engineering practice, on the thick of life, and on erudition of educated society [1]. Vernadsky believed that without studying the "science roots" we won't manage to understand the very core of the modern scientific phenomena, and our image of science genesis, probably, would be false. Vernadsky's biocosmic views are expressed in his understanding of man as planetary force, and in his understanding of the evolution of society as a global-historical regular process with its natural foundations [2].

On the simplest level, Science is a body of empirical, theoretical, and practical knowledge about the natural world, produced by scientists who emphasize the observation, explanation, and prediction of real world phenomena. Historiography of science, in contrast, often draws on the historical methods of both intellectual history and social history. The HS is marked by a chain of advances in technology and knowledge that have always complemented each other. Technological innovations bring about new discoveries and are bred by other discoveries, which inspire new possibilities and approaches to longstanding science issues. Thus, HS studies the historical development of science and scientific knowledge, including both the natural sciences, social sciences, and engineering.

2 "History of Science" as a Part of Modern Science

It was in the seventeenth century that modern science was really born, and the world began to be examined more closely, using instruments such as the telescope, microscope, clock and barometer. It was also at this time that scientific laws started to be put forward for such phenomena as gravity and the way that the volume, pressure and temperature of a gas are related. In the eighteenth century much of basic biology and chemistry was developed as part of the Age of Enlightenment.

Since the mid-nineteenth century, ideas about the HS have been tied to important philosophical and practical questions, such as whether scientific conclusions should be regarded as progressing towards truth, and whether freedom is important for scientific research. Put broadly, the field as a whole examines the entire spectrum of human experience relating to science and technology, and how our understanding of that experience has changed over time.

We might deem the HS of great value when we consider its role in representing the activity of science to wider publics in order to guarantee support for scientific work. Indeed writing about the relationship of science to society has been one of the major directions of the HS, as well as describing the history of scientific ideas and theories, the path of scientific discovery, the growth of scientific knowledge, and the progress of rationality as embodied in and guaranteed by the methods, practices and institutions of scientific inquiry.

The discipline "History of science" concerns the history of the way nature has been manipulated, modelled and understood by different societies. However the study of the HS has had great effects on the philosophy of science, conceptions of the role of science in society, and scientific policy.

In 1931, the Second International Congress of the History of Science was convened in London. The paper entitled "The Social and Economic Roots of Newton's Principia" delivered by Boris Hessen, the member of the Soviet delegation, was devoted to the physical scientist's most famous work, where the author asserted that Newton's work was inspired by his economic status and context, that the"Principia" was little more than the solution of technical problems of the bourgeoisie. Hessen's thesis had a wide effect in Western history of science as its focus on the relationship between society and science was, in its time, seen as novel and inspiring. It was a challenge to the notion that the history of science was the history of individual genius in action. This method of doing the history of science became known as "externalism". It is an approach which eschews the notion that the history of science is the development of pure thought over time, and one idea leading to another which could exist at any place, at any time, if only given the right geniuses.

During the twentieth century, a number of interdisciplinary scientific fields have emerged. For example, "Communication studies" combines animal communication, information theory, marketing, public relations, telecommunications and other forms of communication; "Computer science", built upon a foundation of theoretical linguistics, discrete mathematics, and electrical engineering, studies the nature and limits of computation. Subfields include computability, computational complexity, database design, computer networking, artificial intelligence, and the design of computer hardware. One area in which advances in computing have contributed to more general scientific development is by facilitating large-scale archiving of scientific data. Contemporary computer science typically distinguishes itself by emphasising mathematical 'theory' in contrast to the practical emphasis of software engineering. Materials science has its roots in metallurgy, mineralogy, and crystallography. It combines chemistry, physics, and several engineering disciplines. The field studies metals, ceramics, plastics, semiconductors, and composite materials.

The HS is no longer an isolated discipline but it has become fully a part of history proper. It gives us an opportunity to trace all the changes in various branches and foresee the further possible development of each scientific direction. Thus, looking for 200-year's history of engineering we can observe a chain of its development: practical mechanics—applied mechanics—theory of mechanisms and machines, from great rise to the period of recession [5]. Today, rapid development of scientific-and-technical progress brought to a totally new level of machine design based on the advancing progress of absolutely new branches of science, such as nanotechnology, biomechanics, composite materials, laser technology, and etc.

3 Ethics and Aesthetics in Engineering

The discipline goal is introducing students to the history of engineering and also to identify important, not strictly technical, aspects of engineering and engineering design, such as the engineering ethics, engineering aesthetics; the philosophers' and artists' view of engineering. Thus, ethics and aesthetics studying became an integral part of the HS researches.

Modern society is influenced by revolutionary and accelerated changes in science and technology that challenge in different ways some basic implicit and explicit moral assumptions and legal norms. This is one main reason why there is a growing need for ethical reflection on morality and law. Each new invention or new machine may be both a good thing and a bad thing. That is why ethics should be applied on all stages of research, such as planning, conducting and evaluating a research project in engineering in particular. Otherwise, emergence of threat to all existence of mankind and even emergence of world accident is inevitable. For example, we want to make sure no one creates a robot that goes out and hurts people.

The HS course helps to provide appropriate training and oversight in the responsible and ethical conduct of research to undergraduate students, graduate students, and postdoctoral researchers participating in any research projects. It also serves for carrying out some basic Ethical Standards where the researchers should avoid any risk of considerably harming people or the environment; should not plagiarize the work of others; not commit science fraud, falsify research or otherwise conduct scientific misconduct. We never have to forget the awful use of nuclear energy when scientific achievements were used for destruction of Human beings.

4 What is TechnoSphere?

Starting as an online digital environment launched on September 1, 1995 and hosted on a computer at a UK university TechnoSphere today is the part of the physical environment affected through building or modification by humans. TechnoSphere is a groundbreaking study that distinguishes the natural time of the cosmos from artificial mechanistic time.

Dr. Peter Haff of Duke, chair of the Nicholas School's Earth and Ocean Sciences Division, defines it as the total environment in which we live. It is the sum of the natural, human, and technological systems and processes that surround us. It includes for example forest ecosystems, animals and machines, nanotechnology, the internet, highways, medical systems, power grids, human populations, political parties, governments and bureaucracies, robots and religions and their interactions with each other. In an age in which both the level and acceleration of technology are high, understanding and living with our "environment" can only mean understanding and living with neoenvironment.

The HS plays an important role while explaining what TechnoSphere is. Historians of science consider a link "TechnoSphere—biosphere" not as really natural, but investigate the developing world of machines and technology in frames of a system "Human being (society)—TechnoSphere". From this point of view the TechnoSphere as artificial material organization is to be built into natural processes.

5 HS Research Object

In the end of XX—beginning of XXI century the object of historical and scientific research has changed: now not only discovery and development of new ideas is fixed, but also history of gaining collective scientific knowledge at various historical stages is being studied. Time of the lone inventors has passed. They were replaced by numerous groups of scientists organized into major research schools, centers, foundations, and labs.

The HS is ought to answer, why some scientists (scientific schools) were able to make discoveries, and others were not, even though they were in a favorable environment. We have to analyze errors, misconceptions, and impasses, besides the cases of success.

For example, one striking example of the science importance can be learned from the Moscow plumbing breakdown. Pipes laid under the freezing ground used to break very often. A well-known scientist in the field of mechanics, Professor Zhukovsky, solved the problem of water hammer. It helped to accurately identify places of pipe breaks and to eliminate the cause by replacing the slide gate flaps with ventils for closing pipe dampers slowly. There are many similar examples of science payback.

6 HS Goals in Technical Sciences

One of the main goals of technical universities, announced in their Regulation [3, 4], is "protection and extension of moral, cultural and scientific values of society; forming in students the sense of patriotism, love and respect for people, traditions of their country, reputation of university". Therefore, the main objectives [6] of the HS in technical sciences are:

1. Systematization and classification of existing knowledge, analysis and synthesis of scientific and technical facts.
2. Broadening the research base.
3. Analyzing the Science role in cultural development of society.
4. Improvement of methodology, comprehension clarification, criticism of conceptual description models and explanation of scientific and technical knowledge, increased use of experimental testing and mathematical description of the historical knowledge, expanding the use of modern computer technologies.

5. Identification and analysis of the "science-technology" relations as a single system, study of science and technology development laws, evaluation of the ideas and hypotheses effectiveness based on the scientific knowledge of the past, forecasting the future.
6. Pedagogical goal: studying of HS enriches educational process, teaches to reason and to disclose theoretical principles, promotes the assimilation of theoretical knowledge, increases their visibility.
7. Educational goal: examples of life and work of great scientists draws students' attention to the moral aspect of their work, forms basic scientific ethics, such as proving the truth, prohibition of plagiarism, pursuit of knowledge, social responsibility to mankind and society, devotion to science.
8. Studying technical science and engineering development features in different regions and countries at different times.

7 HS Teaching at Technical Universities

The HS course has to include a lecture part and a practical training. Lectures should be full of the operation of numerous machines description, so it is advisable to supplement lectures with visual range: animated diagrams, excerpts from documentary films, computer animations, videos. Experience has shown that students understand material much better if they pass a test or write a library-research paper at the end of the course.

Also the HS course should include some independent research: finding archival material in libraries, museums, interviews with participants of historical events or members of design teams. Students are encouraged to make movies, computer models, and working models.

Due to the fact that the HS should be a part of general engineering training, it should be lectured by a staff from a general Engineering Department such as Theory of Mechanisms and Machines.

For example, TMM Department at MSTU named after Bauman has a unique mechanism collection-museum, founded in 1845 by I. P. Balashev and later developed by A. Yershov, F. Orlov and other Russian engineers. It has over 500 educational demonstration models and a training lab where students learn experimental methods of mechanisms investigation. The Museum and the lab allow studying the HS on specific examples.

SPbSTU also has gained a considerable experience in HS teaching, by TMM Department in particular [7]. Some students get very interested in topics they study for their course works, so they create their own layouts and models. This is encouraged by adding additional points to the final grade. Models may be more or less complex. However even a simple model requires an ability to work not only with his hands, but also by his head.

Practical classes (seminars), held for an academic group, allow focusing on studying technology related to the specialty of the group (robotics, industrial machines, printing machines, etc.). In the final sessions coursework can be introduced and graded. Discussions tend to be more active, if the pre-designated "official opponents" are chosen. After the course work is completed, it is saved on a CD and sent to an "official opponent", who is chosen among the students. The opponent makes a response, which is read at the presentation and put with the files of the course work.

At the beginning of a semester it is important to determine deadlines of the course work and set a date for the presentation for each student. It makes it easier for students if they make a weekly schedule for the task. Practice has shown that, despite a large work performed by a student, delays are not common. After presentation all the documents must be turned in to the archive.

8 Conclusion

The main goal of HS teaching at technical universities is the showing that science and technology are driving forces of history. They combine general philosophical, specific technical and scientific, historical and cultural origins. The HS course helps to structure the information field of various disciplines' achievements, problems that affect development of human society, and thus see relationships between different issues that are being solved by experts of various specialties. Nowadays, global problems solving is impossible without a broad interdisciplinary approach. Therefore, the HS course is equally important for students in both areas of human sciences and technical education.

Experience has shown that the HS as an independent discipline needs a few general comments:

- Firstly, the HS course should be introduced in the curriculum of technical universities because it successfully combines many disciplines, greatly increasing their effectiveness.
- Secondly, the HS course helps to learn morality and to understand the place and role of the humanities, both in education and in all scientific and technological culture of the third millennium.

References

1. Vernandsky, V.I.: History of science selected works, pp. 45–46. Nauka, Moscow (1981)
2. Vernandsky, V.I.: History of science selected works. Kant and science, Moscow (1981)
3. Bauman Moscow State Technical University Regulations, Mascow (2011)
4. Saint Petersburg State Polytechnical University Regulations, Saint Petersburg (2012)
5. Frolov, K.V.: Scientific and information support of the engineering and machine reliability development. Engineering and machine reliability problems, vol. 1, pp. 3–13. Nauka, Mascow (2005)

6. Egorova, O.V.: History of science and technology in technical sciences. Science and education development. Higher education teaching problems, In: Proceedings of International Scientific conference (9–10, Nov 2010), vol. II, pp. 36–41. MSOU, Mascow (2011)
7. Evgrafov, A.N.: Modern technologies usage in teaching Industrial products design and modeling. Cultural heritage and socio-cultural perspectives on development of Russia: Daily life culture. In: Russian Scientific conference materials, pp. 274–278. USTU-UPI, Yekaterinburg (2010)

Research Lead Student Projects on Multi-Disciplinary Optomechatronics with Applications in Biomedical Imaging

V.-F. Duma, M. L. Negrutiu, C. Sinescu, I. Voiculescu, E. Miutescu, A. Burlea, M. Vlascici and Adrian Gh. Podoleanu

Abstract The education side of an optomechatronics consortium of academic and industrial partners is presented. The consortium covers a variety of disciplines including mechanical design, theory of mechanisms, mechatronics, as well as optical engineering with applications in specific biomedical fields, such as dentistry and gastroenterology. The five teams involved in the project compound senior and young researchers, including graduate and undergraduate students). The contribution of each partner is presented, with examples of specific methodology in conducting student projects on subjects inspired by research.

Keywords Higher education · Students · Research · mechanisms · Mechatronics · Optomechatronics.

V.-F. Duma (✉)
Aurel Vlaicu University of Arad, Arad, Romania
e-mail: duma.virgil@osamember.org

M. L. Negrutiu · C. Sinescu
Victor Babes Medicine and Pharmacy University of Timisoara, Timisoara, Romania

I. Voiculescu
City University of New York, New York, USA

E. Miutescu · A. Burlea
Arad County Emergency University Hospital, Arad, Romania

M. Vlascici
S.C. Inteliform S.R.L., Timisoara, Romania

A. Gh. Podoleanu
University of Kent at Canterbury, Canterbury, UK

J. C. García-Prada and C. Castejón (eds.), *New Trends in Educational Activity in the Field of Mechanism and Machine Theory*, Mechanisms and Machine Science 19,
DOI: 10.1007/978-3-319-01836-2_16,

1 Introduction

Multi- and inter-disciplinary directions are essential in the rapidly evolving landscape of Science today. Rarely work can still be limited to a single, classical field like Mechanical (ME), Electrical (EE), or Optical Engineering (OE). A blend is usually necessary between different fields, while Information Technology (IT) has become a must in all of them. Another tendency is to reunite science and engineering fields with biological and medical areas, by example, and these inter-disciplinary domains stand for some of the most demanding applications today. In this paper we briefly present some of the main directions of research approached in our Consortium—related to the competences of each partner—with specific emphasis on the student-related activities [6], at both UG (undergraduate) and PG (postgraduate) level. Some of these directions lead to training of Master and Doctoral students.

2 Research Consortium

Optomechatronics is a multi-disciplinary field that unites OE and Mechatronics—the latter being itself a blend of ME, EE, and IT. Biomedical Imaging (BI) is an inter-disciplinary field that uses Photonics and other Physics fields for advanced investigation and diagnostics techniques in Biology and Medicine.

The Consortium will mainly be concentrated in the future projected metropolitan area to unite the two closest cities in Western Romania, Arad and Timisoara, and includes several foreign partners. The structure of this Consortium, designed as a future Photonics Pole is shown in Fig. 1. It includes the two universities that are at the core of this structure: the *Aurel* Vlaicu University of Arad (UAVA)—represented

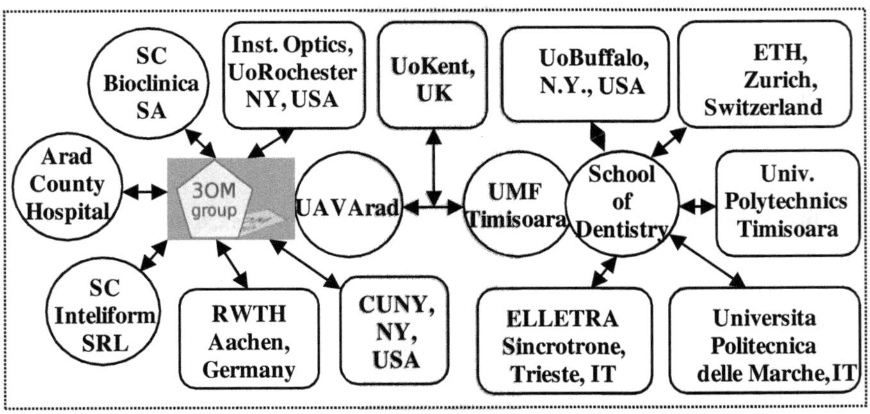

Fig. 1 The members of our consortium—future Photonics Pole in Western Romania (marked with *circles*) and some of our main collaborators (marked with *rectangles*)

by the 3OM Optomechatronics Group [21]—and the Victor Babes University of Medicine and Pharmacy (UMF) Timisoara—represented by the Imaging Group of the Medical School of Dentistry. These two institutions were first put in contact by our mutual collaborator, the Applied Optics Group (AOG) at the University of Kent (UoKent), Canterbury, UK. The Consortium also includes the Arad County (Clinical University Emergency) Hospital, S. C. Bioclinica S. A. (the largest array of medical analysis laboratories in Western Romania), and S. C. Inteliform S. R. L. (a dynamic SME focused on Mechatronics) from Timisoara. In Fig. 1 these partners—of our current Partnership project [22]—are marked with circles, and our most important partners from abroad with rectangles.

The activity in our Project [22] involves at present 30+ researchers, with a balance between senior and young researchers—the latter including both UG and PG students. We have presented in [6] some of our activities involving especially UGs in the 3OM Group (UAVA). In the following, research activities with student involvement (in both PG training and UG teaching) will be briefly pointed out for all the academic partners involved.

3 Research Topics Involving Students at the Partner Institutions

(i) **The 3OM Optomechatronics Group at UAVArad**. Our activity is centered on OME devices and systems. The former direction includes scanners (polygonal, Risley prisms—presented in [7], and galvanometer-based (GSs)), choppers, attenuators. The latter direction includes optical metrology, radiometry and colorimetry, as well as current developments in Optical Coherence Tomography (OCT).

One of the main research avenue is represented by GSs [11], for which we have demonstrated [4] the optimal input functions (in order to obtain the highest possible duty cycle η, i.e. time efficiency of the device): linear on the active portions plus polygonal for the flyback portions—in contrast to the literature [11], where it was considered that the linear plus sinusoidal functions are best. Our experimental investigations of the most used GS input also demonstrated [5] that the triangular functions are better than sinusoidal and sawtooth—in terms of both η and artifact-free images in OCT. With millimeter depths and micrometer resolution capability, this interferometric technique [8] allows for in vivo real-time imaging of human tissues, even of parts of the body that are affected by rapid involuntary movements, such as the eye—as investigations of the retina by example is a main avenue in OCT [13]. Our current project is focused on the development of two different OCT systems and of different modules for these systems, including GS-based handheld probes and endoscopic miniature scanning heads.

Dual axis GSs scanners consisting of orthogonal devices are traditionally used in OCT (but also in other BI techniques, such as confocal microscopy) for the lateral scan of the sample. GS modulation was also developed for other purposes in OCT, such as scanning delay Fourier Domain (FD) OCT.

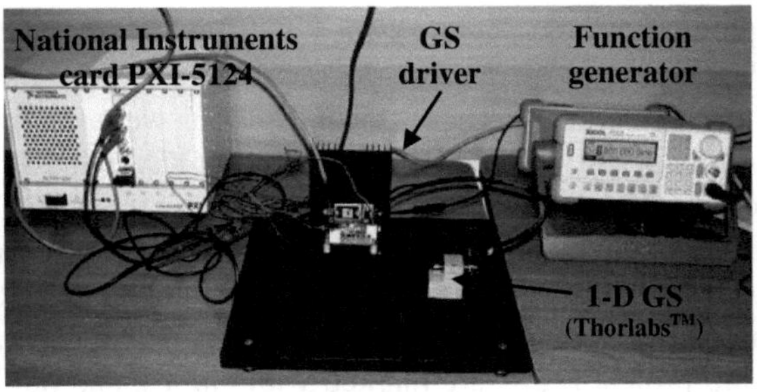

Fig. 2 An experimental setup for the study of GSs (for both UG and PG students)

A FD OCT system was used in [5] to obtain experimentally the necessary rules-of-thumb to optimally drive the GS for different regimes required, including scan frequencies and amplitudes. This research, carried out in the frame of a Fulbright grant at The Institute of Optics of the UoRochester, NY, also involved a postdoc and a PhD student. Several students are now working in GS modulation (as well as on other types of scanners) in the 3OM Group. This activity is also linked to the essential collaboration on OCT with the AOG at the UoKent (Fig. 1), included for the design and manufacturing of the handheld heads with GSs for OCT.

Scanners were thus introduced as one of the most important topics in Instrumentation (lecture and labs, 4th year EE UGs). An experimental stand used in the lab for the study of GS modulation is presented in Fig. 2.

(ii) **The Imaging Group at the School of Dentistry of the UMF Timisoara** is focused on different BI techniques, which include OCT and CT (Computer Tomography). Both TD (Time Domain) and SD (Spectral Domain) OCT systems developed with the AOG, UoKent are in use—mainly for dental studies [16], both of hard and soft tissue in the oral cavity. An example of such a system and an investigation are presented in Fig. 3. These OCT systems were used in different interdisciplinary research concerning noninvasive diagnostics in neurology, dermatology, and cardiology—in order to develop new techniques dedicated for each medical specialty. The results obtained with OCT were validated by microscopy, histology, MicroCT, confocal microscopy, SEM and TEM investigations.

(iii) **The Applied Optics Group (AOG) of the University of Kent** (UoKent), UK [23], led by Professor A. Gh. Podoleanu has built up an international reputation based on its extensive publication record in the areas of theoretical and applied optics, and has attracted considerable funding from the EPSRC, BBSRC, Leverhulme Trust, ERC, EC, NIHR, Pfizer, OTI Inc. Canada, New York Eye and Ear Infirmary (NYEEI) and Innovative Small Instruments.

The AOG has pioneered *en-face* OCT technology in 1996 and has produced the first OCT *en-face* images from the retina, has researched and protected the generation of a combined OCT/confocal image [13], generated 3D images from retina and skin, devised multi-interferometer configurations and special modulators to collect simultaneously images from different depths in the tissue and introduced the concept

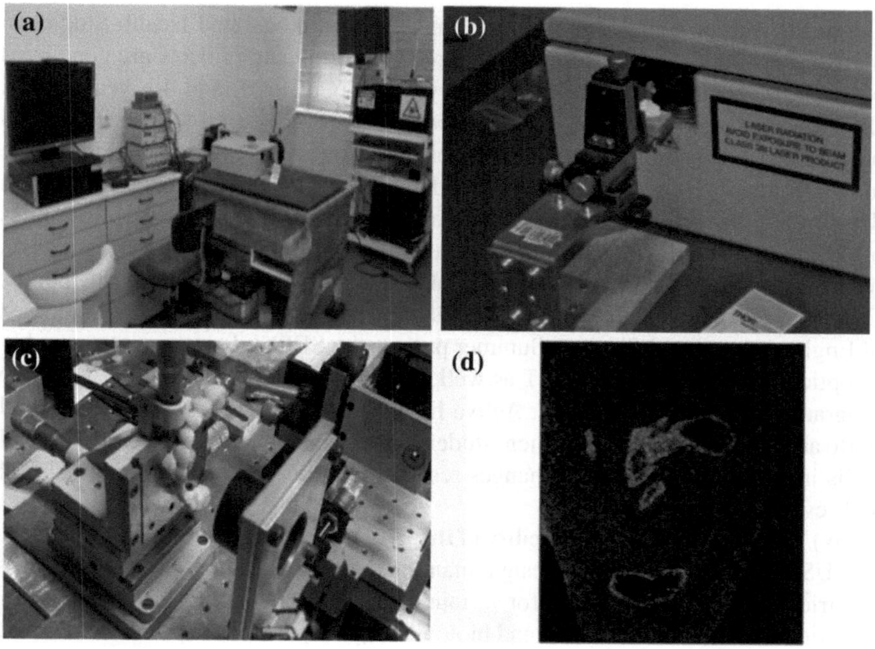

Fig. 3 **a** TD OCT system working at 1300 nm and SD OCT system from the UMF School of Dentistry; **b** TD OCT evaluation of a scaffold from a rat bone; **c** TD OCT for dental studies; **d** macro-retentions from the infrastructure of a metallic-polymeric dental crown

of OCT imaging with adjustable depth resolution. AOG also performed theoretical studies on the performances of OCT systems. First images of eye with pathologies using *en-face* OCT have been reported. AOG has also developed the first instrument combining OCT with ICG fluorescence for imaging the retina in collaboration with New York Eye and Ear Infirmary and Ophthalmic Technology Inc. Combination of three technologies in one system, OCT, scanning laser ophthalmoscopy and adaptive optics and highest limit in vivo obtained in *en-face* images as thin as 3 μm were reported. Resources were oriented towards extending the expertise acquired from ophthalmology to cell imaging and OCT applicability to image embryos has been investigated [10] and more recently, to OCT endoscopy [2]. Expertise was extended to dentistry [16], to imaging of basal cell carcinoma and to art conservation [9]. Other current research focuses on a novel solution to reject the effect of mirror terms in FD-OCT [14], to eliminate the speckle in OCT, on 3D imaging using multiple paths configurations [15, 20], on coherence gated wavefront sensors [19] and on speeding up the acquisition based on graphics cards [17].

Professor Podoleanu has been involved in supervision of postdoctoral researchers constantly from 1999, with a number fluctuating, up to 5/year and of PhD students, with a number up to 8/year. He has created a new lecture course with support from the EC, *Biomedical Optics*, taught to the PhD students in the School of Physical Sciences and to the Master students in the Kent Institute of Medical and Health Studies in the UoKent (2006–2010). Research inspired undergraduate projects are conducted by the AOG with students from the 3rd year MSc (PH600) and 4th year MPhys (PH700) in the School of Physical Sciences. They cover several aspects of applied optics, such as designing hand held probes for imaging the eye, for imaging in the mouth, for endoscopy and addressing fundamental limits in the technology, such as devising simple methods for linearisation of data before FFT [12]. An applied optics project is conducted with students from the 3rd year Forensic Science programs on investigating the OCT as an anti-spoof tool in imaging fingerprint at security points. Other projects are conducted with students from other universities in the South East of England, supported for short Summer projects by SEPNET, embracing assembly of optical configurations for OCT as well as devising customised LabView Matlab programmes for OCT interfaces. Active links with Universities in Brno, Lille and Porto are supporting work placement students on 3–10 months who develop practical skills in an optics lab. Such exchanges resulted in published results in journals [1] and several conferences.

(iv) **The Nanofabrication Facility of the City University of New York** (CUNY), NY, USA is specialized on the design, manufacturing and testing of MEMS (Micro-Electrical-Mechanical Systems) for various applications, which include biochips for detection of hazardous chemical and biological agents, medicine (pressure sensors), communication (circuits), or micro-fluidics [18]. Its collaboration with the Consortium is related to the manufacturing and testing of certain types of endoscopes for OCT. While these activities are developed with the implication of the PhD students employed in the Facility, in Fig. 4 a "Class on a chip" MEMS structure [24] is presented, as used for the UGs in lab works (where visiting students are also received on a regular basis, especially from France).

The Fundamentals of Mechatronics course at CUNY (ME 31100) includes [3] by example lab works for: stress and strain measurements, temperature measurements

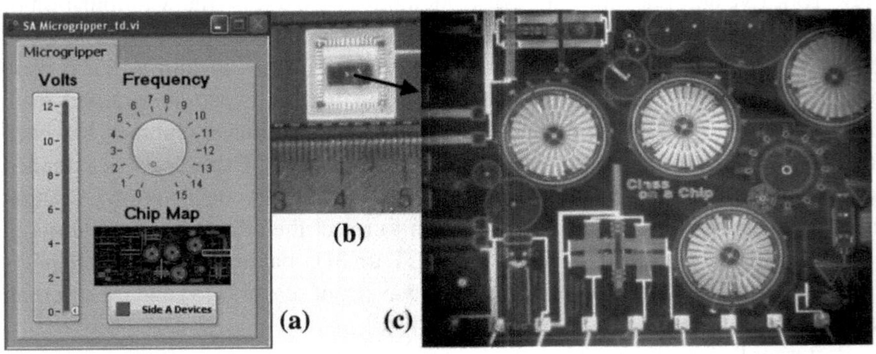

Fig. 4 Class on a chip [24] study of MEMS with UGs at CUNY Nanofabrication Facility: **a** virtual interface; **b** actual size of the chip; **c** partial view of the chip—under microscope (including micromotor and gears, microgrippers, comb drive, parallel and series chevrons)

Fig. 5 Examples of mechatronic systems developed at S.C. Inteliform S.R.L. Timisoara

with thermocouples, optoelectronic monitoring the speed of bodies during free fall, and study of mechanical vibration using four transducers (capacitive, velocity, piezoelectric accelerometer, and linear variable differential transformer).

(v) **Arad County (Clinical University Emergency) Hospital** is represented in the Consortium by the Department of Gastroenterology and by the Department of Pathological Anatomy. They apply the OCT systems and the different handheld and endoscope probes developed in the project for endoscopy—in gastroenterology, as well as in colonoscopy. The biological samples are studied first *ex vivo*, than in vivo as by the end of the project the technique is expected to move into the clinic. Several research objectives are targeted, including polypoid lesions (to be either sampled or removed endoscopically) and Barrett's esophagus. Established procedures like Chromoendoscopy, Autofluorescence Imaging, Ultra-magnifying Endocytoscopy or Confocal Laser Endomicroscopy reduce unnecessary biopsies or resections, and decrease the risk of endoscopic complications. Unfortunately they are expensive, labor intensive, time consuming, and need experienced MDs. OCT will serve to differentiate between neoplastic and non-neoplastic lesions, and to improve the accuracy of the differential diagnosis (e.g., colon adenoma).

(vi) **S.C. Inteliform S.R.L. Timisoara** is a SME specialized on ME and Mechatronics (but also with capabilities in OE)—for the design, simulations and manufacturing of both pieces and systems. Its role in the Consortium is to achieve the technological design and the manufacturing of prototypes, as well as to provide in part the necessary technological transfer of the results of the research (Fig. 5).

4 Conclusions

We presented our consortium, achieved for our current Partnership project [22], but also for the Photonics Pole we are establishing in Western Romania. Some of the main directions of research of the partners are pointed out, as well as the translation of the expertise gained through research in the educational process in the universities involved—for both UG and PG students. This presentation is focused on the multi-disciplinary specific of our work in Optomechatronics. This work includes classical fields like instrumentation and automation, but also photonics, biomedical imaging—with a focus on OCT, applied to non invasive high resolution imaging in a variety of medical fields (e.g., gastroenterology and dentistry).

Acknowledgments This work was supported by a grant of the Romanian Authority for Scientific Research, CNDI project PN-II-PT-PCCA-2011-3.2-1682. A. Podoleanu acknowledges the ERC 7th Framework Programme, Advanced Grant 'COGATIMABIO' 249889 and the NIHR Biomedical Research Centre at Moorfields Eye Hospital NHS Foundation Trust and UCL Institute of Ophthalmology.

References

1. Bouchal, P., Bradu, A., Podoleanu, A.Gh.: Gabor fusion technique in a Talbot bands optical coherence tomography system. Opt. Express **20**(5), 5368–5383 (2012)
2. Cernat, R., Tatla, T.S., Pang, J., Tadrous, P.J., Bradu, A., Dobre, G., Gelikonov, G., Gelikonov, V., Podoleanu, A.Gh.: Dual instrument for in vivo and ex vivo OCT imaging in an ENT department. Biomed. Opt. Express **3**, 3346–3356 (2012)
3. Delale, F., Liaw, B.M., Jiji, L.M., Voiculescu, I., Yu, H.: Infusion of emerging technologies and new teaching methods into the mechanical engineering curriculum at the city college of New York. Adv. Eng. Educ. **2**(4), 1–36 (2011)
4. Duma, V.F.: Optimal scanning function of a galvanometer scanner for an increased duty cycle. Opt. Eng. **49**(10), 103001 (2010)
5. Duma, V.F., Lee, K.-S., Meemon, P., Rolland, J.P.: Experimental investigations of the scanning functions of galvanometer-based scanners with applications in OCT. Appl. Opt. **50**(29), 5735–5749 (2011)
6. Duma, V.F.: Teaching Mechanisms: from classical to hands-on-experiments and research-oriented. In: Pisla, D., Ceccarelli, M., Husty, M., Corves, B. (Eds.) Mechanisms and Machine Science, Vol. 5, Part 8, pp. 493–501, Springer, Berlin, (2010)
7. Duma, V.F., et al.: Optomechatronics applications of the theory of mechanisms with active student involvement in research. In: Proceeding of ISEMMS (in print) (2013)
8. Huang, D., Swanson, E.A., Lin, C.P., Schuman, J.S., Stinson, W.G., Chang, W., Hee, M.R., Flotte, T., Gregory, K., Puliafito, C.A., Fujimoto, J.G.: Optical coherence tomography. Science **254**(5035), 1178–1181 (1991)
9. Liang, H., Cid, M.G., Cucu, R.G., Dobre, G.M., Pedro, J., Saunders, D.: En-face optical coherence tomography: a novel application of non-invasive imaging to art conservation. Opt. Express **13**(16), 6133–6144 (2005)
10. Ma, L., Bradu, A., Bloor, J.W.: Arrhythmia caused by a drosophila tropomyosin mutation is revealed using a novel optical coherence tomography instrument. PLoS **5**(12), 1–8 (2010)
11. Montague, J.: Scanners—galvanometric and resonant. In: Driggers, R. G. (ed.) Encyclopedia of Optical Engineering, pp. 2465–2487, Taylor and Francis, New York (2003)
12. Payne, A., Podoleanu, A.Gh.: Direct electronic linearization for camera-based spectral domain optical coherence tomography. Opt. Lett. **37**(12), 2424–2426 (2012)
13. Podoleanu, A.Gh., Rosen, R.B.: Combinations of techniques in imaging the retina with high resolution. Prog. Retinal Eye Res. **27**, 464–499 (2008)
14. Podoleanu, A.Gh.: Unique interpretation of Talbot bands and Fourier domain white light interferometry. Opt. Express **15**(15), 9867–9876 (2007)
15. Rogers, J.A., Bradu, A., Podoleanu, A.Gh.: Polarization maintaining multiple-depth en face optical coherence tomography system using active re-circulation loops in the non-stationary state. Opt. Express **20**, 29196–29209 (2012)

16. Sinescu, C., Negrutiu, M.L., Todea, C., Balabuc, C., Filip, L., Rominu, R., Bradu, A., Hughes, M.: Quality assessment of dental treatments using en-face optical coherence tomography. J. of Biomed. Opt. **13**(05), 054065 (2008)
17. Van der Jeught, S., Bradu, A., Podoleanu, A.Gh.: Real-time resampling in Fourier domain OCT using a graphics processing unit. JBO Letters **15**, 030511 (2010)
18. Voiculescu, I., Zaghloul, M.E., McGill, R.A., Houser, E.J., Fedder, G.K.: Electrostatically actuated resonant microcantilever beam in CMOS technology for the detection of chemical weapons. IEEE Sens. J. **5**(4), 641–647 (2005)
19. Wang, J., Podoleanu, A.Gh.: Demonstration of real-time depth-resolved Shack-Hartmann measurements. Opt. Letters **37**(23), 4862–4864 (2012)
20. Zurauskas, M., Rogers, J., Podoleanu, A.Gh.: Simultaneous multiple-depths en-face optical coherence tomography using multiple signal excitation of acousto-optic deflectors. Opt. Express **21**, 1925–1936 (2013)
21. http://www.3omgroup.appspot.com/
22. http://3omgroup-optomechatronics.site88.net/
23. http://www.kent.ac.uk/physical-sciences/research/aog/index.html
24. http://classonachip.com/

Analysis of Dynamic Systems Using Bond Graph and SIMULINK

José Antonio Calvo, Carolina Álvarez-Caldas and José Luis San Román

Abstract The aim of this paper is presents an educational application, developed in MATLAB, which allows the engineering students to learn easily and quickly about dynamic systems behaviour through Bond Graph method. This application uses the SIMULINK library of MATLAB, which has proven to be an excellent choice in order to implement and solve the dynamic equations involved. Based on block diagram of SIMULINK, the different "bonds" of Bond Graph can be integrated as SIMULINK blocks in order to generate the dynamic model. As an example, a simple model are analysed through this application.

Keywords Bond graph · Simulink · Dynamic systems · Simulation

1 Introduction

The dynamic systems analysis, very common in engineering studies, is relatively simple when the steady state behaviour is analyzed, or when the system has few degrees of freedom. However, for complex systems, the problem can be highly complicated and the classic way to establish the behaviour equations becomes inadequate.

In most of the cases, the main concern of engineering students is to establish the mathematical model that represents the dynamic behaviour of the system and how the different parameters influence the system behaviour, because the equations that

J. A. Calvo (✉)
Mechanical Engineering Department, Universidad Carlos III de Madrid, Madrid, Spain
e-mail: jacalvo@ing.uc3m.es

C. Álvarez Caldas
e-mail: calvarez@ing.uc3m.es

J. L. S. Román
Universidad Carlos III de Madrid, Madrid, Spain
e-mail: jlsanro@ing.uc3m.es

J. C. García-Prada and C. Castejón (eds.), *New Trends in Educational Activity in the Field of Mechanism and Machine Theory*, Mechanisms and Machine Science 19,
DOI: 10.1007/978-3-319-01836-2_17,

represent the dynamic of the system are usually partial differential equations, whose solutions require deep mathematical knowledge that in most cases is not available for the students.

The Bond Graph technique [1] is extraordinarily useful to overcome these difficulties. Bond Graph is a simple and effective method to set out the differential equations of any dynamic system independently of the physical field analyzed. Bond Graph provides a common model for a wide range of systems ranging from the usual Electric, Mechanics, Hydraulic, Thermals, etc., or combinations of them.

Depcik et al. [2] compare different software and language options, which are available to build models of dynamic systems. They establish that MATLABTM and Simulink^{TM} [4] might be the best choice if the teacher wishes to collaborate with students because engineering students are typically familiar with MATLAB.

In this paper, we present an application, developed in Simulink library, which allows the engineering students to learn easily and quickly about dynamic systems behaviour through Bond Graph method.

After a brief introduction to the Bond Graph method, it will be explained how Simulink can be applied to improve the method, transforming the Bond Graph graphical diagram to a Simulink block diagram. Next, a mechanical example of the application will be presented in order to demonstrate the benefits of the method. Finally, the results of the dynamic response will be analyzed using the various tools provided by Simulink.

2 Bond Graph Theory

There exists a vast literature about Bond Graph method and its applications to analyze dynamic systems such Vera et al. [7], Thoma [6], Margolis [5] and Karnopp [3]. In this paper, the fundamentals of the Bond Graph theory will be presented in order to understand how to implement a model in BONDSYM but the present work's focus is not to explain the Bond Graph method in detail.

Energy is a basic commodity in a system. It flows in from one or more sources, is temporarily stored in system components or partially dissipated in resistances such as heat, and it finally arrives at "sinks" or "loads" where it produces some desired effects. Power is the rate at which energy flows and is a scalar with no direction.

Bond Graph represents this power flow between two systems. This flow is symbolized through an arrow as Fig. 1 illustrates. Unfortunately, it is difficult to measure power directly, and engineers prefer to work with two temporary variables called "flow" and "effort". Depending on the physical environment, these variables have different values. In electrical networks, flow represents the "current" and effort the "voltage", whereas in mechanical linkages, flow represents the "velocity" and effort the "force". The product of both temporary variables is power as Eq. 1 shows in the case of a mechanical system:

$$Power = f(t) \cdot e(t) = v(t) \cdot F(t) \tag{1}$$

On each bond, one of the variables constitutes the cause and the other the effect. The relationship is indicated on the bond, so that it is possible to deduce that quantity in one part of the graph contributes to another quantity somewhere else. Effort and flow causalities always act in opposite directions in a bond. Power bonds may join at two kinds of junctions: a "zero" junction and a "one" junction. In a "zero" junction, the flow and the efforts satisfy the following expressions:

$$\begin{aligned}\sum_n Flow_{input\_n} &= \sum_m Flow_{output\_m} \\ Effort_{input\_1} &= Effort_{input\_2} = \cdots = Effort_{input\_n} \\ &= Effort_{output\_1} = \cdots = Effort_{output\_m}\end{aligned} \tag{2}$$

This corresponds to a node in an electrical circuit (where Kirchhoff's current law applies). In a "one" junction, the flow and the efforts satisfy the following expressions:

$$\begin{aligned}\sum_n Effort_{input\_n} &= \sum_m Effort_{output\_m} \\ Flow_{input\_1} &= Flow_{input\_2} = \cdots = Flow_{input\_n} \\ &= Flow_{output\_1} = \cdots = Flow_{output\_m}\end{aligned} \tag{3}$$

This corresponds to a force balance at a mass in a mechanical system. As an example of a "one" junction, consider a resistor in series. In junction, the principle of energy conservation is assumed, no loss is allowed. There are a set of elements used to model the real systems such as:

- **Resistor**: This element represents a situation where a loss of energy appears (Electrical resistor, mechanical damper, Coulomb frictions, etc.).
- **Compliance**: This element represents the situation where storage of energy appears (electrical capacitors, mechanical springs, etc.).
- **Inertia**: This element represents the relationship between the "flow" and momentum (electrical coil, mass, moment of inertia, etc.).
- **Sources**: This element represents energy sources. There are two kinds of sources: flow source and effort source.
- **Transformer**: A transformer adds no power but transforms it, like an electrical transformer or a lever.
- **Gyrator**: A gyrator relates flow to effort. It does not add power.

3 Mechanical Example of One Degree of Freedom

In order to understand the Bond Graph method used on a simple mechanical model. Figure 1 illustrates a single degree of freedom system composed of a rigid body that can only move up and down, and represented by a mass "m_1", a spring (K_1) and

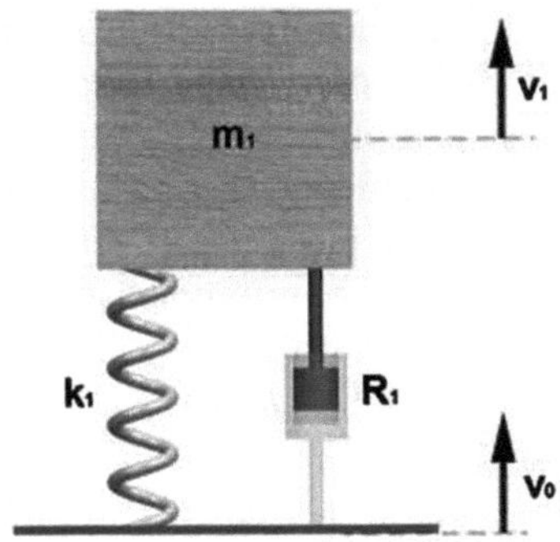

Fig. 1 One degree of freedom system

a damper (R_1). The source of energy is a known velocity of the ground $V_0(t)$. The variable is the mass velocity $V_1(t)$.

First step is representing the real system as a Bond Graph diagram. Usually the Bond Graph starts with sources, in this case a flow source (ground velocity). The next step is to think about the power flow through the system components. The power flows by the effort and flow variables. As Fig. 1 illustrates, the spring and damper are moved at a velocity that is the difference between the mass velocity $V_1(t)$ and the ground velocity $V_0(t)$. This means that a "zero" junction must be used as Fig. 2 illustrates.

The spring is represented as a compliance bond and the damper as a resistance bond. Both components move at the same velocity, but each one needs a different force to move at this velocity. This means that a "one" junction must be used as Fig. 2 illustrates. Finally, the mass is represented as an inertia bond that moved at $V_1(t)$ velocity, and as the system is under the gravitational force of earth an effort source must be used.

Bond Graph method calculates the flow and effort on each bond, and uses the displacement associated to compliance bond (X) and the momentum associated to Inertia bond (P) as system variables.

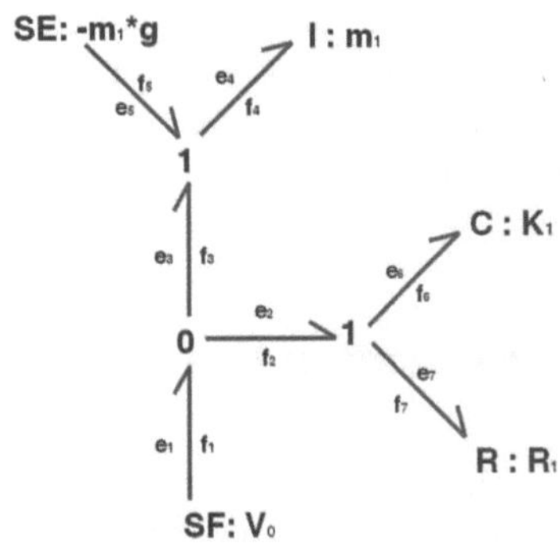

Fig. 2 Bond graph of the one DOF system

Two coupled differential equations represent the system's dynamic behaviour as Eqs. 4 and 5 show. This is obtained by assuming that the variation of the momentum is equal to effort on inertia bond, and that the variation of the displacement is the flow in compliance bond.

$$\frac{dX(t)}{dt} = f_7 = V_1(t) - \frac{P(t)}{m_2} \tag{4}$$

$$\frac{dP(t)}{dt} = e_4 = K_2 \cdot X(t) + \left(V_1(t) - \frac{P(t)}{m_2}\right) \cdot R - m_2 \cdot g \tag{5}$$

4 Simulink Application

Simulink is an environment for multi domain simulation and Model-Based Design for dynamic and embedded systems. It provides an interactive graphical environment and a customizable set of block libraries that allow designing, simulating, implementing, and testing a variety of time-varying systems. All these features of the program allow generating a block model suitable to represent the Bond Graph model. The Simulink block model will solve the equations and allows the user to analyze the dynamic behaviour of the system.

This section explains how to use the benefits of Simulink to set up and solve the equations that manage system behaviour. The procedure consist in convert the real model into a Bond Graph model and then translate it to the Simulink block diagrams, as can be seen in Fig. 3.

These blocks of Simulink will be used to represent the action that occurs in each element of the Bond Graph model. Figure 4 illustrates the Simulink block diagram of the one degree of freedom system.

The Flow source (ground velocity) is connected to a sum block (one junction) that calculates the difference of velocities between the ground (V_0) and the mass (V_1). This velocity is the input (known flow) for the spring (Compliance Bond) and the damper (Resistor Bond), that move at the same speed.

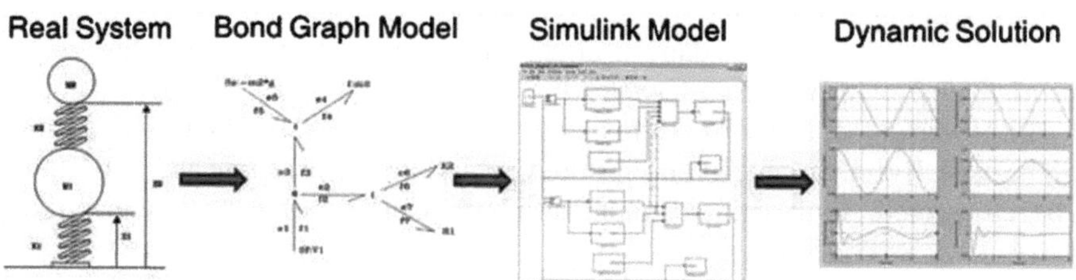

Fig. 3 Procedure to analyze the dynamic behaviour of a system

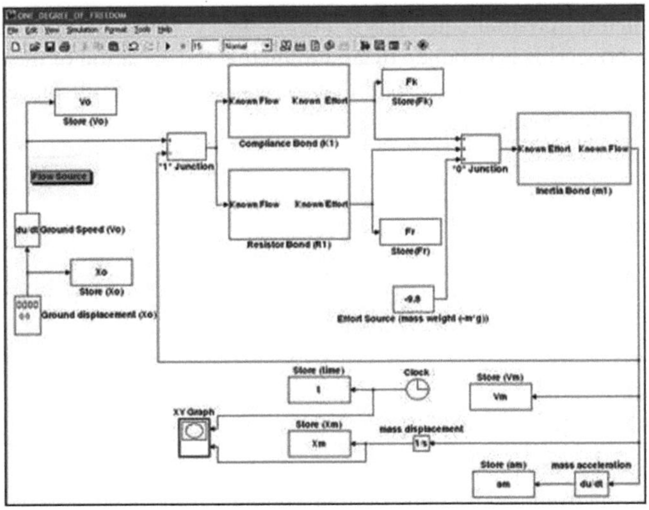

Fig. 4 Simulink one degree of freedom model

They return to the system the spring force (F_K) and the damper force (F_R) that are connected to a sum block (zero junction) to calculate, by adding gravity force (Effort Source), the equilibrium force.

This force will be the input to the Inertia block in order to calculate the velocity of the mass (V_1) that is used as an iterative loop to calculate the difference in velocity between the ground and the mass. Due to the fact that the input to the system must be a flow (ground speed) it is necessary to derivate the ground displacement in order to obtain a flow source.

5 Model Results

In order to simulate the free movement of the system, we assume that the ground is stopped (Vo = 0 m/s) and the spring is not preloaded (Xo = 0 m), this means that the mass will fall on top of the spring and therefore the system moves (this is known as free vibration). Under the damper effect, the system will stop after a few cycles due to the loss of energy in each cycle.

The powerful tool of MATLAB to plot the results, the MATLAB Graphics Editor, will be used. Every variable has been stored in MATLAB workspace and is available to be represented. Numerical and graphical analysis can be done using the capabilities of MATLAB as Fig. 5 shows.

By looking the different graphs of the variables that have been plotted, it is possible to analyze what is happening to the model when the mass can move freely.

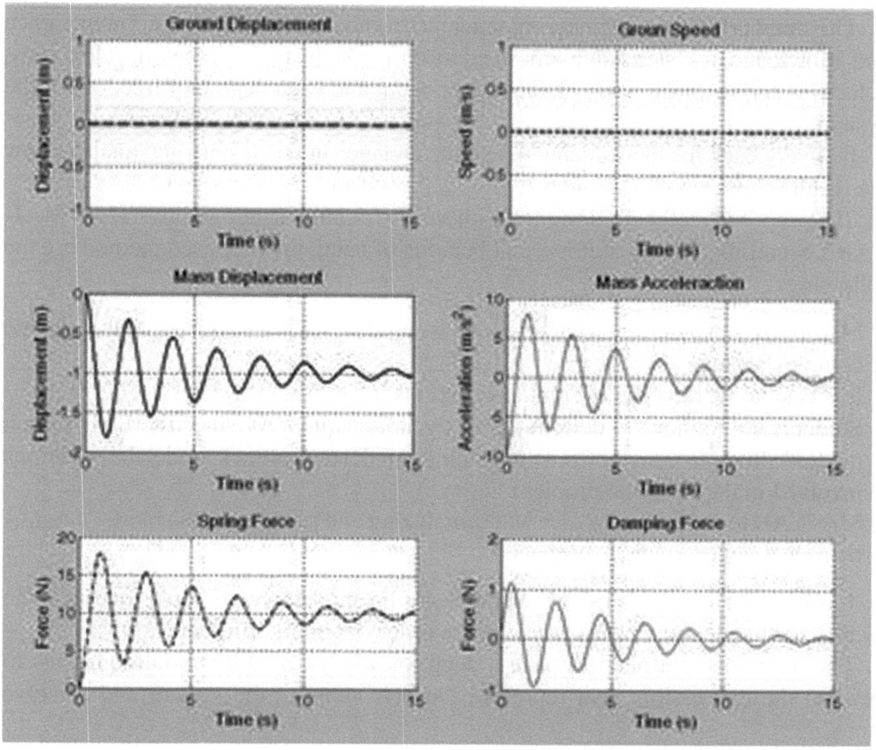

Fig. 5 Results for the one degree of freedom model with no ground displacement

Initially, the mass falls and compresses the spring. Then, when the inertia and gravity forces achieve a value equal to the spring force, the mass returns to the steady state position. This movement is repeated but, as consequence of the damper, the system loses energy and finally stops after few cycles. When the system stops, the spring is loaded with a force equal to the mass weight (9.8 N) so the mass stabilizes at minus one meter from the initial position.

Now, It is very easy for the student to modify the model parameters (mass value, spring rate, damping ratio) and see how the behavior of the system changes, or increase the model complexity by adding of mass, springs and dampers and evaluate the results.

6 Conclusions

This work presents a method to apply the Bond Graph technique to implement and solve the dynamic equations of a dynamic system by taking advantage of MATLAB and Simulink.

This method allows engineering students to quickly and easily gain experience and knowledge in systems dynamics and to learn which are the forces, accelerations, velocities and displacements of each component and each degree of freedom of the system.

It is very easy for the students to test the changes in the system and analyze how the results change.

The proposed method is designed with a user-friendly Windows interface in MATLAB's Simulink. The most important benefits of using the proposed method are the following:

- MATLAB toolboxes and functions can be used, allowing the program to simulate complex systems.
- It is easy to work with differential equations, matrixes and vectors.
- Students do not have to determine the dynamic equations, since the Bond Graph method allows to transfer from the graphics model to the block model the equations involved in the dynamic problem.
- MATLAB tools are easy to use when analyzing and comparing the model's results by numerical or graphics outputs.

It is noticeable the facility to generate complex models from simples ones and the facility to change the model parameters in order to obtain different results.

The user does not need to have a deep knowledge of differential equations to develop the expressions that represent the behaviour of the system and to solve them.

References

1. Blundell, A.: Bond graph for modeling engineering systems, ellis horword publishers (1982)
2. Depcik, C., Assanis, D. N.: Graphical user interfaces in an engineering educational environment. Comput. Appl. Eng. Educ. **13**(1), 48–59 (2005)
3. Karnopp, D.: On the order of a physical model, Trans. ASME J. Syst. Dyn. Meas. Control. **101**(3), 102–118 (1979)
4. The MathWorks, Inc. Creating SIMULINK Models (2000)
5. Margolis, D.: Introduction into bond graph. In: 3rd Seminar on Advanced Vehicle Dynamics Amalfi (1985)
6. Thoma, J.U.: Introduction to bond graph and their applications, Pergamon Oxford, Oxford (1975)
7. Vera, C., Aparicio, F., Felez, J.: Simulación de sistemas dinámicos mediante la técnica del Bond Graph. In: Sección de Publicaciones de la E.T.S. I. Industriales de Madrid, Universidad Politécnica de Madrid (1993)

Teaching Machines Tools Operation in Virtual Laboratories of Engineering Faculties

M. J. Martín, F. Martín, C. Bermudo and L. Sevilla

Abstract The knowledge of the different machines operation and their main mechanism has a great importance in mechanical engineer formation. Thus, in the manufacturing field, it is essential to understand the behaviour and basic principles of the machine tools. This paper presents some experiences developed by the manufacturing teachers of the University of Malaga to facilitate teaching of a high number of common machine-tools in mechanical engineering such as lathes, milling machines, drill press, planers or shapers. For this, a lot of technological data have been compiled to recreate the mechanical configurations, with more than 300 pieces in several cases, resulting in a virtual environment. Actually, the high number of students in the Engineering Faculties, the limited availability of these machines in laboratories and the difficulty to visualize the internal mechanisms show that the virtualization of this machine tools, presented in this work, provide an effective mean to transmit knowledge to the Engineering students, serving as an essential complement in the study of motion transmission mechanisms. In order to achieve greater effectiveness, each education project is enclosed with a guide where students can follow the operation of the machine tools in each virtualization.

Keywords Machine tools · Virtualization · Education · Motion mechanism

---

M. J. Martín (✉) · F. Martín · C. Bermudo · L. Sevilla
University of Malaga, Malaga, Spain
e-mail: mjmartin@uma.es

F. Martín
e-mail: fdmartin@uma.es

C. Bermudo
e-mail: bgamboa@uma.es

L. Sevilla
e-mail: lsevilla@uma.es

J. C. García-Prada and C. Castejón (eds.), *New Trends in Educational Activity in the Field of Mechanism and Machine Theory*, Mechanisms and Machine Science 19,
DOI: 10.1007/978-3-319-01836-2_18,

1 Introduction

Since the signing in 1999 of the Bologna Declaration, Spanish universities are immersed in a process of adaptation to the new university educational system, included within the structures defined for the achievement of a European Higher Education Area (EHEA). This adaptation has stated performances in the regulatory field and in the academic context, as is necessary for achieving a change in attitude in teachers and students about how they conceive the teaching and learning at the university level [1].

In this sense, the use of information and communication technologies (ICT) can be a valuable tool in the face of increasing efficiency in university activities, whether teaching, research or management. Universities have been fully developed initiatives aimed at facilitating the access of the university community to use these technologies (wireless, virtual campus) [2–6].

Currently, the need to provide students with enough practices, that allow access to a real application of a particular field to see applied their knowledge, is one of the bigger problems present at the Faculties of Engineering. In some cases the high number of students, low infrastructure or reduced availability of financial resources, do not allows to achieve the planned objectives. For this reason these information technologies can stand as an efficient way to reach those students practical content that would be impossible to be carried out in person by limited resources. The generation of applications in virtual support, to develop these practical content similar to traditional realization, will achieve a situation of independence of the limitations above-referred.

Thus, the practical realization of virtual experiences by students generates three main benefits. On the one hand, greater flexibility allows to access practices contents throughout their learning period, unlike what happens so far, which is restricted to classroom time. In the background, reducing time-to-face allows increasing the number of additional practices and the access to practices normally not available because the cost of required equipment. Finally, the use of these virtual resources provides greater clarity of presentation of different transmission sets and mechanical elements, allowing a greater understanding of the contents by the student.

Actually, the Department of Manufacturing Engineering of the University of Malaga teaches a large number of practices, being the machining [7] by means of machine tools which presents more complexity, to show the various internal mechanisms that produce and transmit motion the force required to perform cutting operations.

Additionally, the characteristics of implemented and deployed virtualization of practices make a vehicle especially valuable in the case of educational services to disabled students, in the event that they come from motor or auditory limitations, by means of the use of additional access way.

In this article, different virtualizations on machine tools are developed, that altogether form a body with which the student can obtain a comprehensive information on the different mechanisms that enable the various machining operations.

The flexibility and power of this tool, for educational innovation, overcomes some of the limitations that arise frequently in engineering schools, such as the high number of students, the inability to increase the number of groups required practices, limited availability of expensive equipment and sufficiently varied security issues during the initial training stage or space limitations in workshops and laboratories.

2 Methodology

Conducting virtual media practices [8] have been applied in three stages, which are displayed in chronological order. After extensive data collection phase technology with which virtualizations document starts the generation process under virtual media practices through various graphic design software such as *Solid Edge*, *3D Studio MAX*, *All Recorder*, *Adobe Premiere Pro 2.0* and *Pinnacle Studio*.

In the first stage has worked in the *Solid Edge* environment (Fig. 1) [9]. This program allows the modeling of each of the parts of the kinematic chains and movement transmission mechanisms which are subsequently assembled. These pieces are designed and modeled one by one, and then be mounted as if it were the real machine (Fig. 2). The main reason for making the parts through this program is the relative easiness when played in three dimensions all kinds of parts with simple operations and exact dimensions.

Solid Edge modeling package is a plan production and enabling the generation of three-dimensional prototypes. All parts modeled in *Solid Edge* will subsequently exported to *3D Studio Max*, for the most power in the input textures and animation [10, 11] (Fig. 3). In order to export these geometries are used to an intermediate format akin to the two environments, such as *iges* format (file extension *.igs*), and can then easily include any scene in *3D Studio Max*.

The second operation to perform with this program is the animation of the whole. With it gives movement to the elements they need to achieve show displacement powertrains and virtualized different machining processes. Finally, after finishing the process, it comes to rendering the work implemented (Fig. 4), i.e. the transfer of

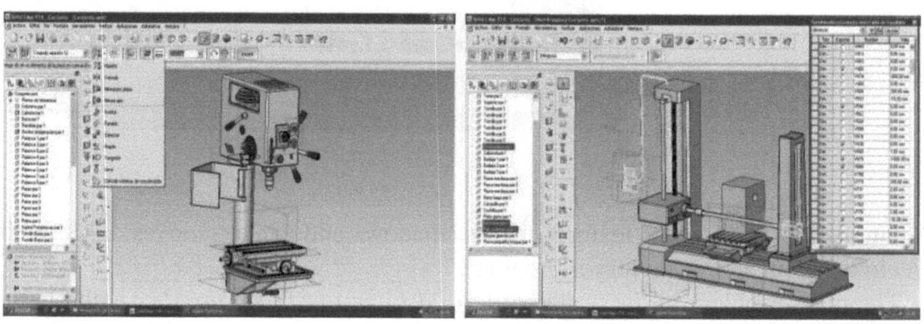

Fig. 1 Solid Edge environment in prototyping

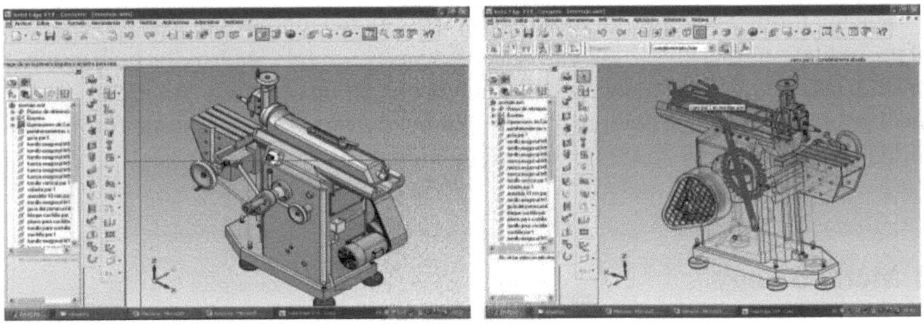

Fig. 2 Transmission mechanisms movements in Solid Edge environment

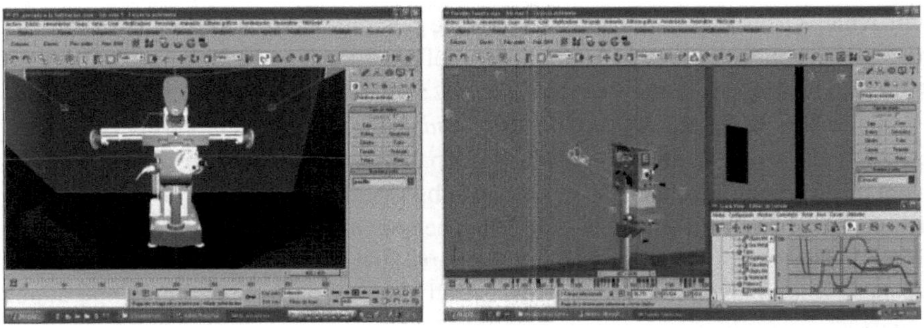

Fig. 3 3D Studio Max environment

Fig. 4 Menu rendering in 3D Studio Max

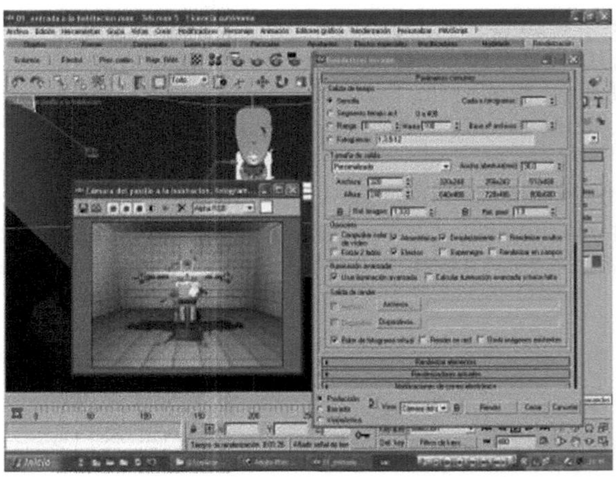

the animation, with all its textures, lights, shadows and camera moves, a file with *.avi* to run it on any video player program.

The main problem in the passage of *Solid Edge* to *3D Studio Max* is the large number of resources consumed by the presentation on the monitor of these surfaces, so working with wireframe. In each case study is presented to solve complex problems

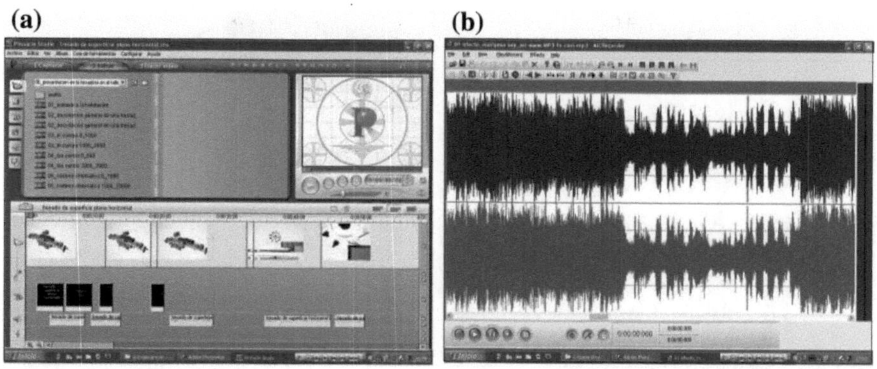

Fig. 5 **a** Pinnacle environment **b** All Recorder

such as the generation of chip type elements or control of fluids, in order to give the greater likelihood virtualization. Using *3D Studio Max* allows easy visualization and representation of the models, and their export and other file formats other than using the program itself.

Due to the required image quality, render times can be very high, so it has resorted to a third video editing program, *Pinnacle Studio* (Fig. 5) [12], as it allows to render the animation videos in parts, significantly reducing render times. After subsequent assembly, the animations are displayed seamlessly, proceeding to turn insert labels and relevant sounds. To reinforce the didactic nature has introduced an audio track mentioning the most important aspects of each machining process. In this audio recording programs are used as alternatives to Pinnacle Studio *All Recorder* (Fig. 5b) [13] that, simply, can record these reviews. Entries are then added and placarded each simulated machining operation.

Finally, the video editing process ends with a new rendering of the images, leading to the final result of the videos that show each of the processes in metal cutting

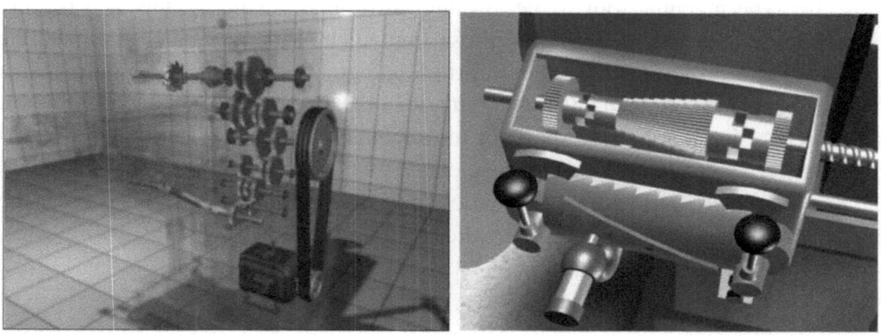

Fig. 6 Adobe Premiere Pro 2.0 environment

Fig. 7 Details of the final result

machining. For the final video used the *Adobe Premiere Pro* (Fig. 6) [14], to unite the various videos and sound.

3 Results

Complementing each virtualized, each educational project has been provided with a guide-manual, with which students can follow the process of manufacture by machining that develops in each practice animation format. The result achieved by applying the methodology set out above, has been conducting a series of practical machining processes in metal cutting, generated in a virtual environment, among others, those for type machine tools lathe, milling, drill press, shaper, planer and boring. By suitable glazes allowed observe those fundamental internal parts and machine tools having a difficult access and constitute integral parts of the kinematic chains responsible for the movements that are made in these machines, such as gearboxes, set of gears, shafts and drive shafts of movements, motion transformation systems rod-crank type, etc. as well as the basic elements of which (main heads) making up the machine itself, and various auxiliaries (indexing table, tailstock) needed for the execution of different machining operations (Fig. 7).

4 Conclusions

Virtual practices allow to complement activities traditionally teaching in person in manufacturing subjects. The flexibility and power of this innovative educational tool shown through a series of examples adequately differentiated, get address some of the most common limitations of engineering schools, such as overcrowded, the inability to increase the number necessary practices group, the availability of a sufficiently varied and expensive equipment, safety problems during the initial training stage or space constraints in workshops and laboratories. The necessary convergence towards the EHEA qualifications of this proposal makes a powerful tool in order to optimize the number of hours available and flexible employment within the framework of the new model of teaching and learning.

References

1. Barros, J., García, L.: The Bologna process as a motivador for change. In: Proceedings of the Informatics Education Europe II, pp. 146–154. Thessaloniki (2007)
2. Balamuralithara, B., Woods, P.C.: Virtual laboratories in engineering education: the simulation lab and remote lab. Comput. Appl. Eng. Educ. **17**(1), 108–118 (2009)
3. Ortega, N., Celaya, A., Plaza, S., Lamikiz, A., Pombo, I., Sánchez, J.A.: Implementation of simulation software for better understanding of manufacturing processes. Mater. Sci. Forum, (New Frontiers in Materials Processing Training and Learning II, LeónStafa-Zurich) **692**, 58–64 (2011)
4. Sanz, A., González, I., Castejón, A.J., Casado, J.L.: Using virtual reality in the teaching of manufacturing processes with material removal in CNC machine-tools. Mater. Sci. Forum, (New Frontiers in Materials Processing Training and Learning II, LeónStafa-Zurich) **692**, 112–119 (2011)
5. Piquero, J.M., Rubio, E.M., Sebastián, M.A.: Computer-aided system for teaching machining with numerical control. Mater. Sci. Forum, (New Frontiers in Materials Processing Training and Learning) **625**, 35–41 (2009)
6. García-Plaza, E., Núñez, P.J., Martín, A.R., Beamud, E.: Virtual machinning applied to the teaching of manufacturing technology. Mater. Sci. Forum, (New Frontiers in Materials Processing Training and Learning II, LeónStafa-Zurich) **692**, 120–127 (2011)
7. Sevilla, L., Martín, M.J., Martín, F., Cano, M.J.: Implantation of virtual practices about materials processing in the manufacturing engineering department of the university of Malaga. Mater. Sci. Forum, (New Frontiers in Materials Processing Training and Learning I—LeónStafa-Zurich, Trans Tech Publications) **625**, 51–59 (2009)
8. Sevilla, J., Martín, M.J., Martín, F.: Integration of virtual manufacturing laboratory of the university of Malaga. Mater. Sci. Forum, (New Frontiers in Materials Processing Training and Learning II - LeónStafa-Zurich) **692**, 65–73 (2011)
9. Solid Edge ST3.: Electronic Data System Corporation (2010)
10. 3d Studio Max 2012.: Autodesk Inc. (2012)
11. Piegl, L.: On NURBS: a survey. IEEE Comp. Graph. Applic. **11**(1), 55–71 (1991)
12. Pinnacle Studio 14.: Pinnacle Systems Incorporated (2010)
13. All Recorder V3 1st.: Benison Software Company (2007)
14. Adobe Premier Pro CS5.: Adobe Systems Incorporated (2009)

Virtual Laboratory Works on Theory of Mechanism and Machine

V. B. Tarabarin, Z. I. Tarabarina and A. G. Feygina

Abstract In a paper is given the description of four virtual laboratory works on the Theory of mechanisms and machines. The description's of works are devoted the research of efficiency of a reducer, dynamic balancing of a rotor, research of influence of cutter tool's parameters on geometry of a involutes tooth and synthesis of four-bars linkages. In works are used the original software and programs MathCAD and AutoCAD.

Keywords Virtual laboratory works · Efficiency · Balancing · Profiling of involutes gear · Synthesis of linkages

1 Introduction

The laboratory works on Theory of machines and mechanisms [1–3] as a part of the course is done to help retain the theory, learn state-of-art methods of experimental research in kinematic and dynamic parameters of machine mechanisms. Performing a laboratory students use physical mechanism models, experimental laboratory stands and other machines provided with different gauges (e.g. speed gauge, displacement gauge, acceleration gauge, force gauge, etc.). During laboratory experimenting kinematic parameters of the points and mechanism chains are defined and investigated, also characteristics of friction process and energy losses in mechanisms can be determined and investigated as well as static and dynamic reactions in kinematic pairs. Both multi-purpose and special computer programs are used for mathematical

V. B. Tarabarin (✉) · Z. I. Tarabarina · A. G. Feygina
Bauman MSTU, Moscow, Russia
e-mail: vtarabarin@gmail.com

Z. T. Tarabarina
e-mail: zitar@mail.ru

G. Feygina
e-mail: a.g.feygina@gmail.com

J. C. García-Prada and C. Castejón (eds.), *New Trends in Educational Activity in the Field of Mechanism and Machine Theory*, Mechanisms and Machine Science 19,
DOI: 10.1007/978-3-319-01836-2_19,

modeling of the processes involved in the research and for the treatment of experiment results.

Though far from perfect but yet high schools are equipped with computer classrooms much better than with laboratory facilities. That's why there is a need to develop virtual laboratory work [4–6].

Such facilities enable students to perform all the phases of a real laboratory work in a dialogue with a computer. Virtual work cannot and should not replace a laboratory work in a real laboratory with all the real devices and facilities there. However, they can be helpful enough under conditions when the real laboratory work is impossible (no facilities at all, no experimental laboratory is available nearby, etc.) while preparations for the laboratory work, exams or final tests. This virtual laboratory work is especially important when it is a distance learning program.

Virtual laboratory work is a complex multimedia program that contains: (1) Instructions for a laboratory work where you can find necessary theoretical information about the subject you learn; (2) Videoclips containing a teacher's explanations or a laboratory assistant's explanations with the following demonstration of the work on a stand; (3) Videoclips of a film from the learning material files which can display existing plant facilities, methods of part treatment and any other learning material that will display a real laboratory work; (4) A computer program that will model the main phases of a laboratory work performance such as preliminary calculations.

All the set of the programs involved is on a hard disc, or a DVD, or on a server available via Internet or local networks. Nevertheless, there are some limitations connected with large volumes of video materials if a virtual laboratory work is done on line [7, 8]. At the TMM Chair at the Bauman University there were developed 8 virtual programs within the time period 1998–2008 [2]. Some of the programs will be considered in brief in the following chapters.

2 Virtual Laboratory Works

2.1 The Laboratory Work "Reducer Efficiency Research"

The task of this work includes: learning the method to determine a reducer efficiency, to define the dependence of the reducer efficiency on the value of resistance moment applied to the output shaft of the reducer, the evaluation of the mathematical model parameters, which would evaluate the dependence of the reducer efficiency on the resistance moment and would determine the value of the resistance moment matching the maximum efficiency value. On performing this laboratory work a student would do all the steps of it in succession. Utility screens of his work are displayed on Fig. 1.

On the first screen there is an introduction. Then you can see the screen formation with the factor change in the experiment (Fig. 1b). Here you can determine the limitations of the experimental setting to change the factors (in this case the engine shaft

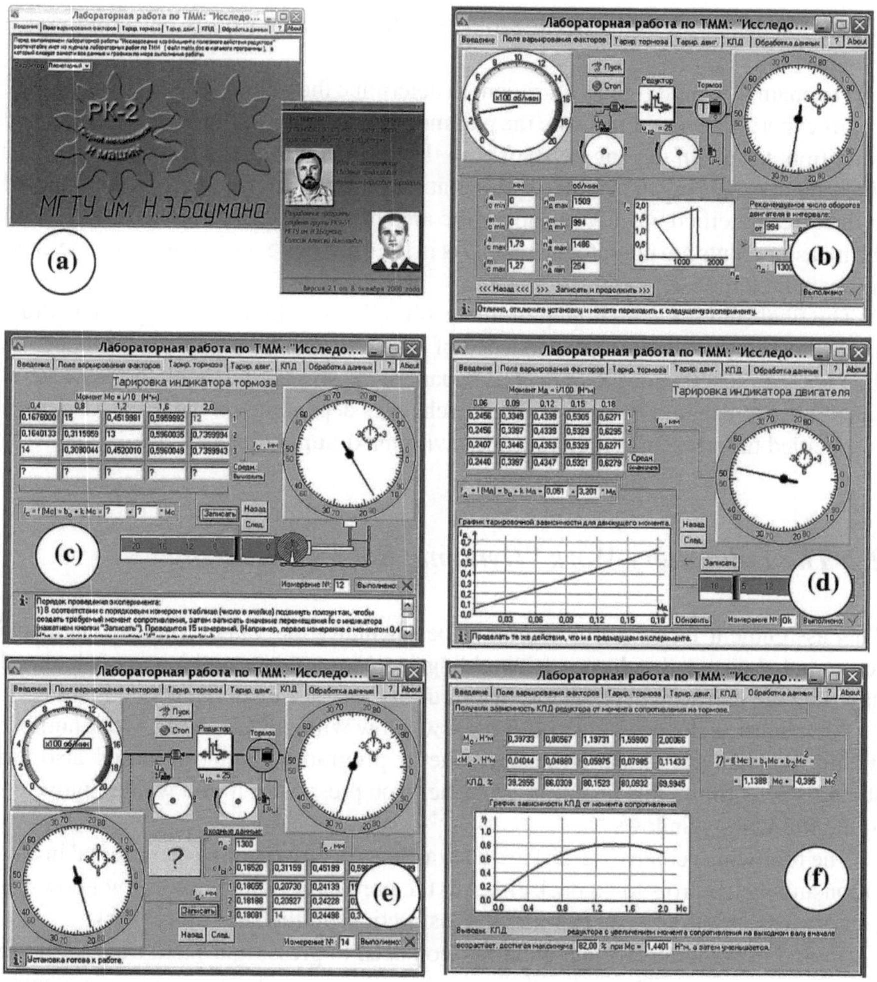

Fig. 1 Screens of the laboratory work "Reducer efficiency research"

frequency and resistance moment of the brakes). After having formed the screen with the factor change you choose an area of investigation within its limits.

The next step in this work is the calibration of the brake indicator (Fig. 1b). In this step the measuring device spring is loaded with fixed torques with the help of a lever and weights. In each phase of load you would record the indicator figures defining the spring deflection. At this load here you'd define it three times.

The third step is the calibration of the engine indicator (Fig. 1c) Here you would conduct the same experiment as mentioned above but you'd measure the torque in the electric engine shaft. According to the results of the experiment you'd draw a calibration chart and would be able to define the coefficients of its linear regressive model.

The main step here is the experiment to determine the reducer efficiency (Fig. 1d). In this experiment you'd measure the parameters necessary for the reducer efficiency definition: torque in the engine shaft with the specified torque in the output reducer shaft. Besides, during the whole experiment the engine shaft frequency is maintained at the level specified when you choose the area of the experiment. The results of all the measurements are put into the table. As in the previous experiments each measure is taken three times.

The final step is the treatment of the results (Fig. 1e). In this phase you can draw a chart where you'll see the dependence of efficiency on the resistance moment and then you can specify coefficients of its square regressive model. With the results you can draw a conclusion how the efficiency changes depending on the load. Also there is specified the resistance moment value with maximum efficiency.

2.2 The Laboratory Work "Dynamic Rotors Balancing"

This assignment means to determine experimentally the values and the angles of rotors unbalance and reduce it with adding correction masses which are calculated based on the experiment results. Program utility screens are demonstrated on Fig. 2.

After starting the program you'll see a window with the laboratory work number within the chair classification and whichever program is being used and also the students performing this task (Fig. 2a). Then you press Start if you want to move on to the following windows.

The following screen (Fig. 2b) includes a photo of an existing experimental installation and the controls to go back either to the previous or the following windows. If you press Instructions, you can find the instructions to the laboratory work.

To find the laboratory main windows you need to press right-hand triangle at the bottom of the screen. The first of the main ones (Fig. 2c) includes a demo of a scheme experimental installation, a correction plane with the controls of the correction mass positions, an indicator to fix maximum amplitude frame, a remote control of main installation elements, an area displaying the frame oscillations and a table to record the experimental amplitude values.

This virtual experiment is conducted as follows: (1) Press the respective Power and Engine buttons; (2) Press the + and − buttons to get the probe mass displaced to a specified position; (3) Press the button Start repetitively until the engine reaches the frequency which exceeds the resonance frequency twice as much. After that the button should be released and the rotor has a running-out. During passing resonance the indicator records the maximum frame amplitude. The value obtained is recorded into the result table if you press the triangle in a table gap (Fig. 2d).

On the window where the data is processed you'll see the value and the phase of the rotor unbalance in the specified correction plane. Also the angles of correction mass placement for its balancing will be seen. And you'll find the vector unbalance

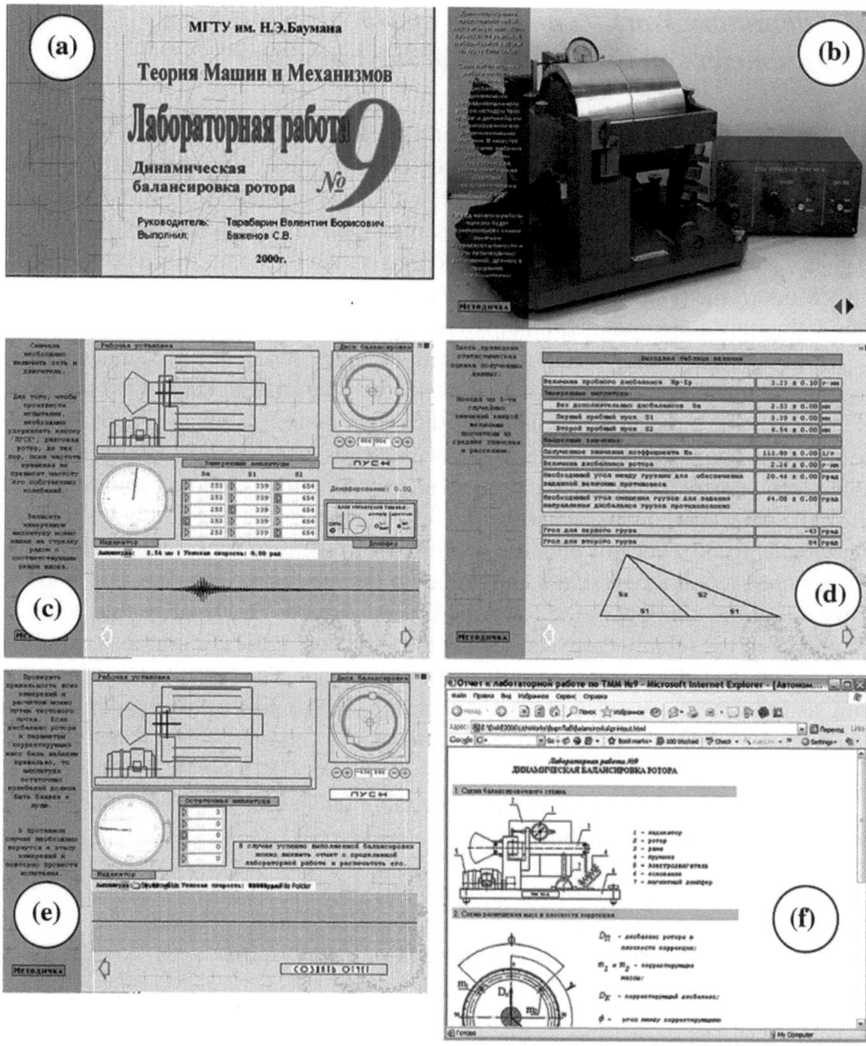

Fig. 2 Screens of the laboratory work "Dynamic rotors balancing"

diagram which would demonstrate the way to measure these values based on three start-ups.

In the next phase of the work the precision of balance is defined and the value of the remaining unbalance is determined (Fig. 2d). If the value of the remaining unbalance meets the demands of the rotor balancing precision, you can shift to the laboratory work report. Here you need to press "Create a report". After checking it you can print it (Fig. 2e).

2.3 Laboratory Work "The Effect of Machine Meshing Parameters on the Geometry of the Cogwheel"

In this work, we model the process of cutting an evolving cogwheel by an instrument with rack producing contour. We also study trimming and sharpening of cogs; and determine the region of existence of a wheel with a regular form of cogs. The sequence of working displays of this assignment is presented in Fig. 3. On the first display (Fig. 3a), we can see a TV set and a video player. We use them to watch the study film "Modern methods of processing the cogwheels" and to record the process of completing this assignment with help of a real machine.

The next display (Fig. 3b) shows the methodology used in this assignment. After reading the methodology and watching the videos the student starts with the work itself. Firstly, he puts the initial data into the program and calculates the main parameters of the cogwheel (Fig. 3c). Second, he proceeds to the section "Practical implementation ", where he models the process of cog trimming for three variants of shift (Fig. 3d). After that, the main geometrical parameters of the cog and the wheel are measured (Fig. 3e). For this end, the measured part of the wheel is zoomed in. The cursor is pointed at the start and end points of the measured distance. The result is displayed in the window in the lower part of the screen. To visualize this, we plot the graphs of the relation between the depth of the cog at the circumferences of the vertices and the splitter from the shift of the instrument. The limiting lines for trimming and sharpening are also plotted. According to these lines, the range of shifts in which the wheel may be processed without trimming and sharpening, is determined. (Fig. 3f).

The conclusion of the work contains the minimal and the maximal values of the shift in the range identified above.

2.4 Laboratory Work "The Synthesis of Four-Link Lever Mechanisms"

Purpose of the work: to choose the optimal value of the generalized coordinate and the size of the links of the mechanism. The optimality is understood in the sense of the proximity of the transmitting function to a particular constant taking the movement of the output link as given.

We consider two mechanisms: crank-slide and four-hinged. On the first stage, the synthesis problem is reduced to the problem of three positions. Further, the size of the mechanism is calculated both analytically and graphically. The input link of the model is fed with three values of the generalized coordinate (the angle coordinate of the link 1). The corresponding coordinates of the output link are measured using these values. Then, the deviations of these positions from the given ones are calculated. According to the values of these deviations, the average error of the model accuracy in representing the given law of motion is determined.

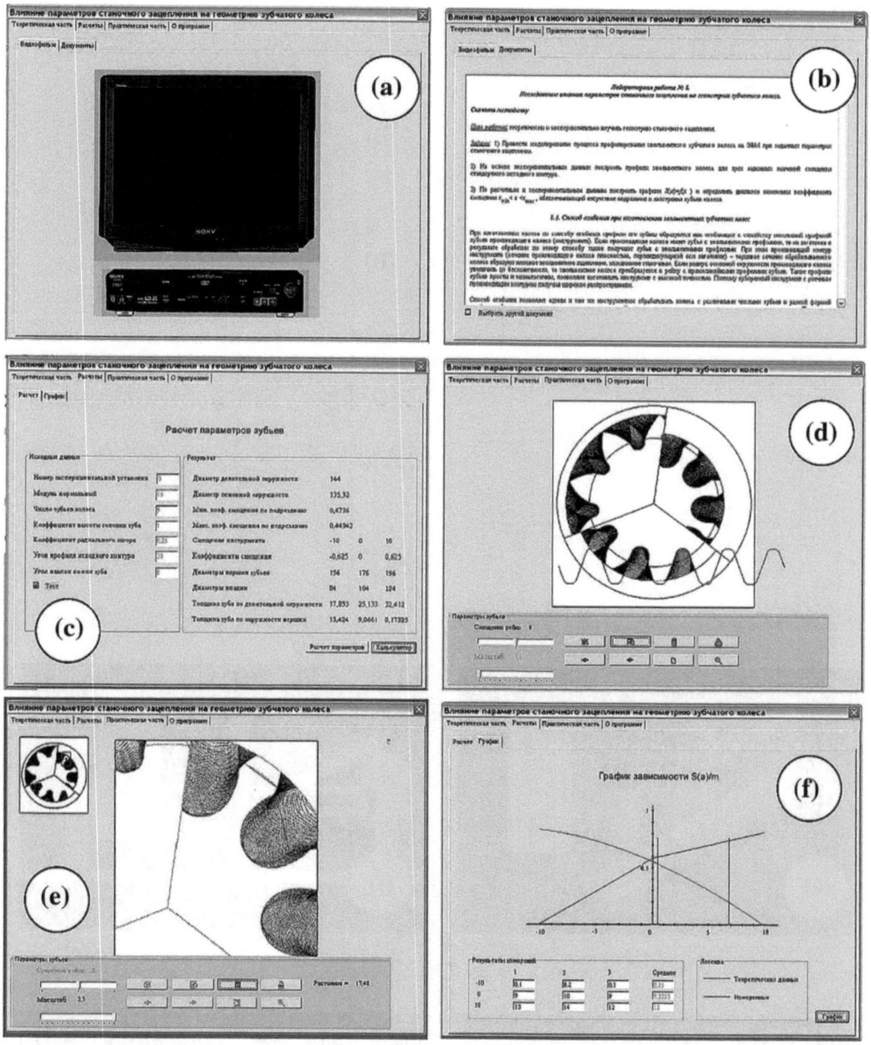

Fig. 3 Screens of the laboratory work "The effect of machine meshing parameters on the geometry of the cogwheel"

The virtual assignment consists of two programs.

- The first one (Fig. 4a–d)—the calculating one. According to a given position function, it randomly chooses three positions of the output link and corresponding changes of the link 1 angle (Fig. 4c). It then analytically calculates the size of the links of the mechanism (Fig. 4d);
- The second one (Fig. 4e–h)—the graphical one. It is developed in *AutoCAD* system to show the 3-D model of the mechanism with alternating lengths of the links.

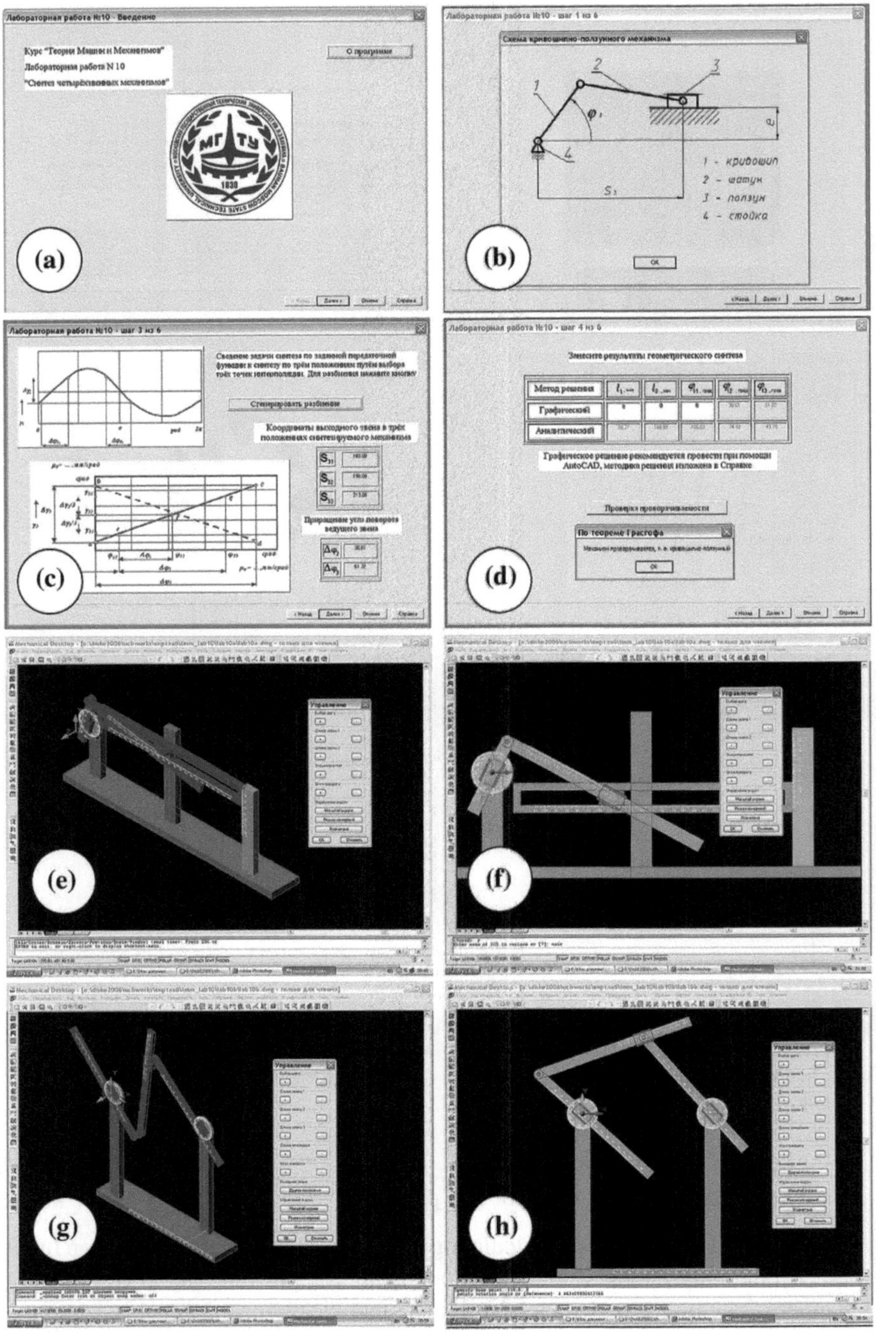

Fig. 4 Screens of the laboratory work "The synthesis of four-link lever mechanisms"

The work is executed in the following order. In the first program, we choose the type of mechanism, input the given values of the initial data, calculate the sizes of the links of the mechanism and check the conditions for existence of cranks. Then, the program *AutoCAD* and *Lisp* file of the corresponding mechanisms are loaded. The sizes of the links and the rotation angle of link 1 are fed into the model with help of a menu. Based on this, the program models the random deviations of the link sizes. In each of the three positions the position of the output link is measured in terms of the model scale. The measurements are repeated three times. The results obtained are then fed into the first program that calculates the average value of the error of the position function.

3 Conclusion

According to the results of the virtual practical assignments in Theory of Mechanism and Machine we can draw the following conclusions:

1. Virtual laboratory works on TMM are the effective modern training tool, both for internal, and for correspondence education.
2. The creating pre-production models of virtual works, have allowed determining both their virtues, and deficiencies that has allowed defining a direction of the further development.

Acknowledgments The authors are grateful to the students specialized in CAD: Bajenov S., Berchun S., Denisov A., and Solosin A., who actively participated in the development of the algorithms and software for the virtual assignments.

References

1. Frolov, K.V., Popov, S.A., Musatov, A.K.: Theory of Mechanisms and Mechanic Machines. Bauman MSTU Press, Moscow (2002). (in Russian)
2. Tarabarin, V., Kuzenkov, V., Fursiak, F.: The Laboratory Practical Work Under the Theory of Mechanisms and Machines. Bauman MSTU Press, Moscow (2002). (in Russian)
3. Tarabarin V., Tarabarina Z. The electronic a learning-methodical complex at the course the Theory of machines and mechanisms. In: Proceeding of the China-Russian Symposium on Problems of Development and Use of the Information Technology in Engineering Education, p. 139. HPU, Harbin, 3 Nov 2007
4. Gamburg K. Virtual bench laboratory works as the innovative form of a context education. The dissertation on competition of a scientific degree of Cand. Tech. Sci.: 13.00.01 Moscow, p. 180 (2006) (in Russian)
5. Petropavlovsk. Virtual laboratory. Use of personal computer at execution laboratory works on general technical disciplines. The electronic resource: http://virtuallab.narod.ru (2003)
6. Nagaeva M.V. Laboratory a practical work with computer support. The electronic resource: Theory of Mechanisms and Machines, http://tmm.spbstu.ru/19/5_nagaeva_19.pdf (2012). (in Russian)

7. Dubinin V., Bannikov S., Jiguliovcev J. Laboratory of the theoretical mechanic on the internet. Creation of system of an open engineering education. BMSTU 2001–2002, Moscow, p. 200 (2003) (in Russian)
8. Zimin A. Automatized laboratory practical works with distant access through the internet. Creation of system of an open engineering education. BMSTU 2001–2002, Moscow, p. 200 (2003) (in Russian)

Interactive Application for Technical Drawing Learning

C. Álvarez-Caldas, J. A. Calvo, A. Quesada and J. L. San Román

Abstract In recent years, we have found that Engineering students do not have the adequate preparation to face the subject of Technical Drawing with guarantees. This forces teachers to find new ways to help students to reach the desired level. The basis of Technical Drawing is the dihedral representation system. So, the present work focuses in helping students to enhance their knowledge about dihedral system as well as their spatial perception. Following the idea of promoting the student's personal work, the designed tool will be an e-learning tool. The proposed application has been programmed with Matlab, and is presented as an executable accessible through the Internet. It allows the student to interact creating, modifying and deleting entities and it provides multiple visualization options.

Keywords e-learning · Dihedral system · Matlab · Web applications

1 Introduction

Descriptive Geometry (DG) is the branch of Geometry that studies the representation of three-dimensional objects on a plane, using systems based on the concept of projecting a point on a plane in order to reduce the three spatial dimensions to the two dimensions of the plane. At present, the basic content of DG (as the basis of the dihedral system) is taught in the last years of pre-university education and in practically all branches of Engineering; it is of vital importance in Design, Mechanical and Civil Engineering [1].

The main purpose of this subject is not only to provide students with theoretical knowledge of Geometryand Drawing, but also to enhance their spatial perception,

C. Álvarez-Caldas (✉) · J. A. Calvo · A. Quesada · J. L. San Román
University Carlos III of Madrid, Madrid, Spain
e-mail: calvarez@ing.uc3m.es

J. C. García-Prada and C. Castejón (eds.), *New Trends in Educational Activity in the Field of Mechanism and Machine Theory*, Mechanisms and Machine Science 19,
DOI: 10.1007/978-3-319-01836-2_20,

one of the seven forms of intelligence and the most essential and vital one in the training of any engineer [1–3].

Although the necessity of having appropriate underpinning knowledge as well as skills and competences that can only be developed through laboratory and workshop practice is clear [4], present syllabuses allow students without this knowledge to arrive to technical degrees.

So the only possible solution is to help the unprepared students to reach the desired level by using distance learning. This is a solution that several teachers are currently using to solve other similar problems [5, 6].

2 Objectives

The main objective of this work is to develop a computer application that can improve the learning process in the basis of the dihedral system.

Such an application is especially designed to help the unprepared students to reach the desired level to face the subject. According to this objective, the application has to fulfill the following requirements:

- Allowing students to learn at their own pace.
- Empowering students to learn through practice.
- Being accessible through the Internet.
- Helping to improve capabilities such as spatial vision.

3 Designed Application

In order to find an application that allows the student not only to understand a set of examples, but also to create new ones, the solution adopted in this paper is to create a new application programmed in Matlab that allows the student to interact creating, modifying and deleting entities. The result will be an executable that will be accessible through the Internet.

The application has to be designed taking into account the general didactical rules, and the particular methodology of the subject as well as the latest empirical studies on e-learning characteristics reported in literature.

The possibilities offered by the application are:

- Generating points in dihedral system.
- Generating lines in dihedral system.
- Generating planes in dihedral system.
- Generating any geometric entity that can be defined as the union of two or more of the previous entities.
- Every created entity, as well as the creation process, can be visualized in dihedral system and in real 3D view, in order to help the student to establish the relation between dihedral and real entities.

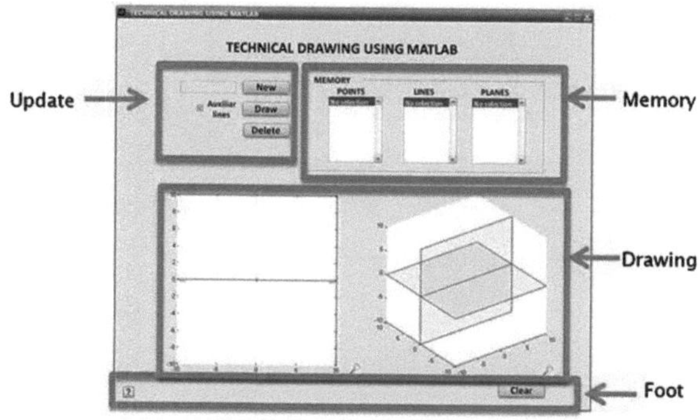

Fig. 1 Main screen of the application

3.1 Workspace

The general aspect of the created application is shown in Fig. 1.

As can be seen, the main screen can be divided in four different areas:

- The Update area: located in the upper left corner of the screen. It includes the controls to create, draw or delete entities.
- The Memory area: right next to the Update area. It shows all the created entities (points, lines, planes…) and it is possible to make a selection to work with it.
- The Drawing area is placed below these two areas, where the bidimensional and tridimensional representations of the entities are shown. These views can be maximized by clicking in the magnifying glasses.
- Finally, at the bottom of the screen, there is the Foot area, where the help button and the clear button—which allows to delete all data in order to start again—are placed.

3.2 Creating Entities

The first action that can be done is to create an entity. It is possible to create points, lines and planes.

Each entity can be created either by introducing data in the textbox or by selecting an already created entity in the listboxes.

Created entities are automatically named, stored in the internal memory of the application, and shown in the corresponding listboxes (points, lines, planes).

In the Drawing area, there are shown two representations of the entities: the dihedral representation and the real 3D view. This allows the students to visualize 3D models together with the corresponding dihedral views, making this representation system much easier to understand and learn for the students.

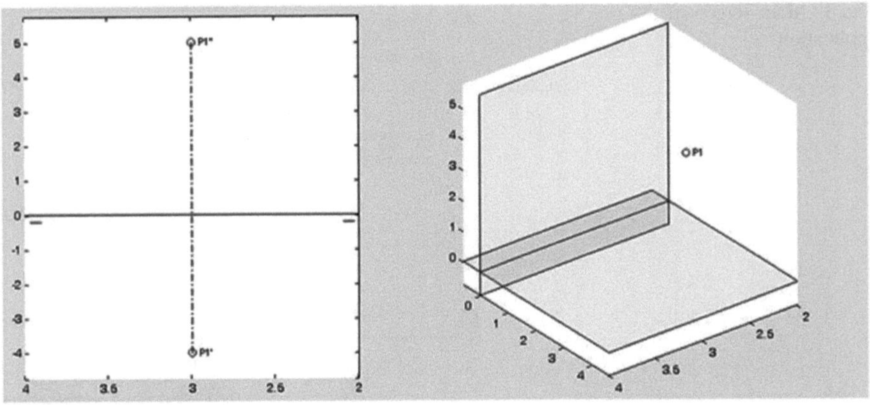

Fig. 2 Representation of a point

The procedure to create an entity is different depending on the kind of entity created.

3.2.1 Creation of a Point

Points can only be created by introducing their coordinates in the textbox placed in the Update area (Fig. 1). Those coordinates must be three real numbers separated by commas (",").

By pressing the New button, the point appears in the Memory area (Fig. 1) and also in the Drawing area (Fig. 2), represented by its projections in dihedral system and by its real position in the 3D view. If the checkbox Auxiliary lines is activated, then the projections of the point are also shown in the 3D view.

3.2.2 Creation of a Line

Lines are created from two already created points. The command can be written in the textbox of the Update area as "RP1,P2" or "rP1,P2". The text recognizer identifies the entity as a line as the program calculates the line defined by these two points, storing it in the Memory area and showing it in the Drawing area. The hidden line policy of the dihedral system has been respected, drawing as solid lines only the part of the line contained in the first quadrant.

The projections of the line are shown in the dihedral view, while only the real line is plot in the 3D view. If the auxiliary lines are activated, projections can be seen also in the 3D view, and so the traces of the line appear in both representations (Fig. 3).

Instead of writing the order in the textbox, it is possible to select the two desired points in the Memory area and just click in the New button.

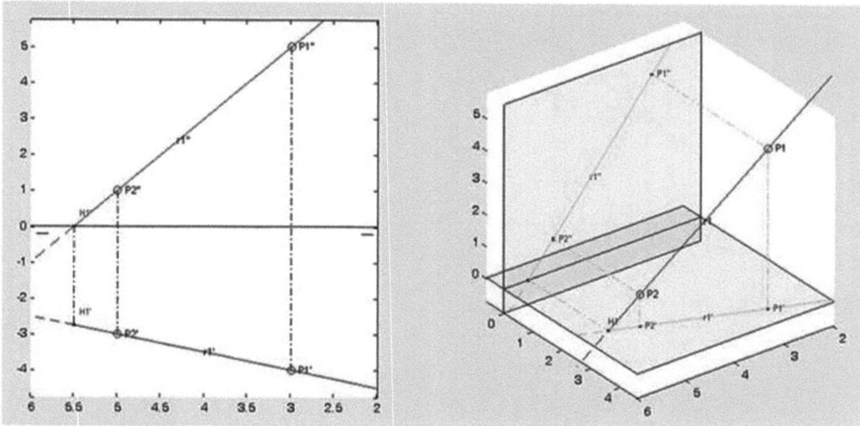

Fig. 3 Representation of a line with its auxiliary lines

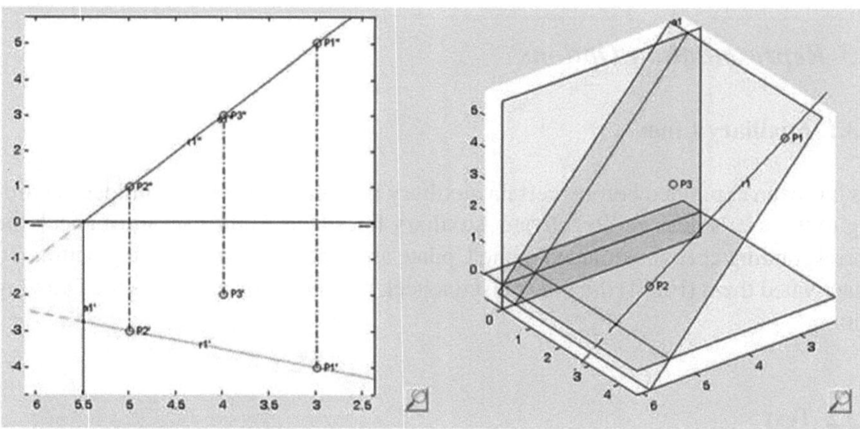

Fig. 4 Representation of a plane

3.2.3 Creation of a Plane

Planes can be created from three entities' sets:

- Three points which are not aligned
- A line and a point which do not belong to the line
- Two parallel lines or two lines with a common point.

In all cases, it is possible to select the desired entities in the Memory area and press the New button or write the corresponding command in the text box of the Update area (e.g. Ar1,P3 creates the plane determined by line 1 and point 3).

The created plane will be shown as in Fig. 4, drawing its traces in both views.

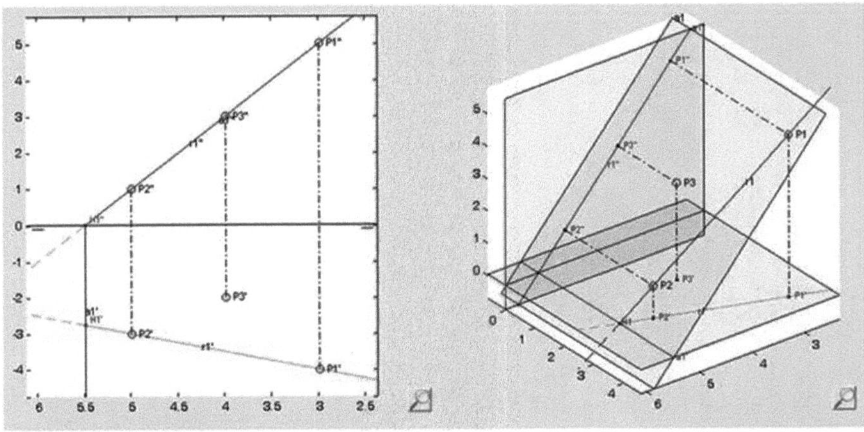

Fig. 5 Plane, line and three points with Auxiliary lines activated

3.3 Representation Options

3.3.1 Auxiliary Lines

As has been explained before, certain auxiliary lines can be shown or hidden according to the user's desires. To activate Auxiliary lines (Fig. 5) the user must check the corresponding checkbox placed in the Update area and press the Drawing button. To deactivated them (Fig. 4) the user must uncheck the checkbox and press the Drawing button.

3.3.2 Text

The names of the entities can be useful sometimes, but they can also complicate the view in certain cases. Because of this, texts can be shown or hidden by pressing the Texts button placed below the Memory area.

3.3.3 Resizing

Any time the program draws a view, the more suitable scale is calculated taking into account the entities that are going to be drawn.

3.3.4 Drawing Selecting Entities

Sometimes it can be useful to represent only some of the created entities. To do it, the user has only to select the points, lines and planes that are going to be drawn in the Memory area and press the Draw button. The application recalculates the scale and offers a clearer representation of the selected entities.

3.3.5 Maximized Views

The most powerful visualization tools of the application are the maximized views.

These views are available by clicking in the maximizing glass buttons placed next to the normal views. The maximized dihedral view is just a zoom of the normal dihedral view, but the maximized 3D view offers the user the possibility of moving 360° around the view by using the mouse those obtaining images as those shown in Fig. 6.

The advantage of this view is that it enables the students to interact with the simulation to conceptualize three dimensional functional relations that are not apparent

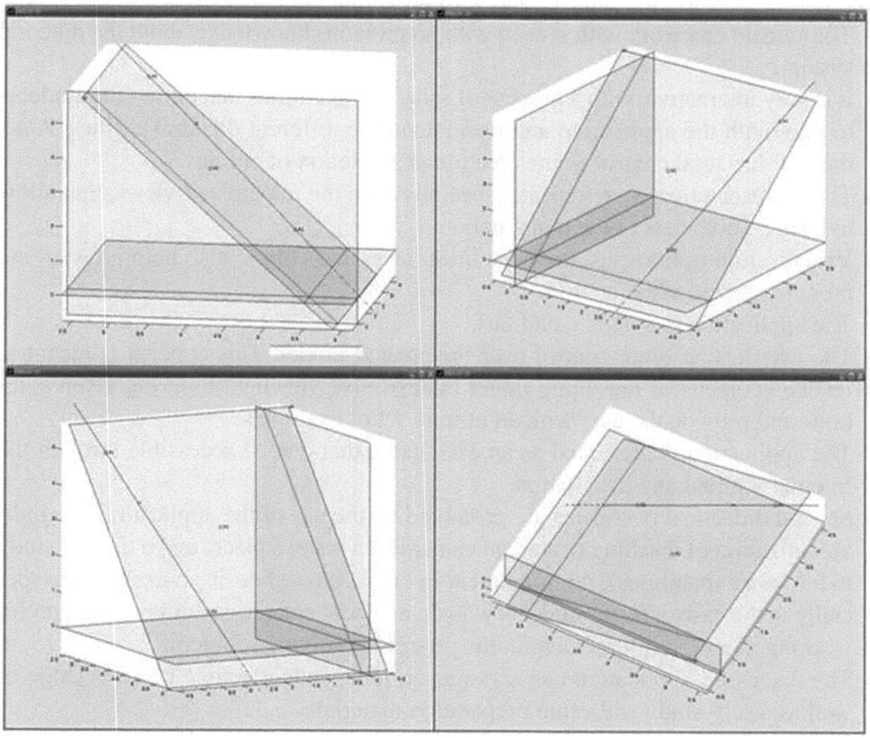

Fig. 6 Different positions of the maximized 3D view

in a less dynamic representation, and it also allows them to visualize 3D models that are difficult to understand in any other way.

3.4 Help File

A help file is always available by pressing the "?" button placed in the bottom of the screen (Fig. 1).

4 Conclusions

The present application is a resource created neither to replace the professor nor the traditional teaching aids, but rather to supplement them by helping the students to understand the basis of the dihedral system and reach the level required to be able to follow the subject.

The main advantages of the proposed application are listed below:

- It is very easy to use, intuitive and with great functionalities.
- The student can work with it even with no previous knowledge about the dihedral system.
- It is very interactive, with a process of selection of entities that allows the students to play with the application and understand the different dihedral entities. Also, the user has total control of the creation and deletion of entities.
- The representation is very clear, especially with the maximized views, that allow to have a 360° view of the construction.
- Visualization options, as auxiliary lines, texts and colors, also help a faster and more organized visualization.
- The application is dynamic and fast.
- The user has absolute control over the created model. This concept is meant to invoke geometrical reasoning rather than passive viewing/observing given solutions and provide the user with an infinite set of examples.
- The application is presented as an executable that can be accessible through the Internet without any installation.
- Several didactical postulates are promoted by the use of the application, eg. individualization of teaching (a student can adjust a pace, a place, and a time of study to his own capabilities), reinforcement in the process of do-it-yourself tasks (specially in the case of those students who do not like "collective thinking" and prefer working alone), reinforcement in the process of solving problems.
- The discussed application can serve as an independent source of knowledge as well as a self–study and exam preparation material.
- Definitively the application can help the students to improve their spatial visualization and their knowledge about the dihedral system.

References

1. García, R., et al. Interactive multimedia animation with macromedia flash in descriptive geometry teaching. Comput and Edu. **49**(3), 615–639 (2007)
2. Contero, M., et al.: Learning support tools for developing spatial abilities in engineering design. Int. J. Eng. Edu. **22**(3), 470–477 (2006)
3. Kaufmann, H., Schmalstieg, D., Wagner, M.: Construct3D: A virtual reality application for mathematics and geometry education. Edu. Info. Tech. **5**(4), 263–276 (2000)
4. Tornincasa, S., Chirone, E.: The Leonardo WEBD project: An example of the WEB3D technology applications for distance training and learning. In: XIV Congreso Nacional de Ingeniería Gráfica. Santander, Spain (2002)
5. Jones, B.W.: Teaching computer graphics in the online environment. In: 59th Annual Midyear Conference of the Engineering Design Graphics Division of the American Society for Engineering, Education. Williamsburg, USA (2004)
6. Rojas Sola, J.I., Internet: La nueva puerta del siglo XXI aplicada a la enseñanza de la geometría descriptiva. Revista Mágina. **10** 197–203 (2002)

Part IV
Mechanical Engineering Education: New Trends

Comparison of Geometry Software for the Analysis in Mechanism Theory

S. Kurtenbach, I. Prause, C. Weigel and B. Corves

Abstract Using geometry software within the process of mechanism development has not yet become common practice in mechanical engineering, even though in the early design stage, in which these tools are often applied, geometry software tools have a huge advantage. At this time the mechanism is not completely designed but consists of its kinematic parameters exclusively. This enables a broad analysis of a quickly generated mechanism respectively a basic synthesis for certain links of this mechanism far away from the first implementation in a CAD-system. In several mechanism lectures at IGM students were invited to familiarize with geometry software thus learning about the development of a mechanism right from the beginning. For this task the interactive geometry software Cinderella© is applied, representing a freeware tool provided by Springer©. But since there is a wide variety of geometry software with large differences in properties, dimensions, graphic user interface, ease of use etc., this paper presents a comprehensive comparison of common geometry software used in many different disciplines of education and later on in working life. This requires a global research into geometry software applied within the process of mechanism development. Subsequently this very long list can be split up into different groups possessing similar characteristics. Based on these characteristics an objective comparison and evaluation will be carried out. The evaluation criteria can be extracted from different show cases where a certain synthesis or analysis procedure is performed. Finally, the result of this paper is a recommendation which

S. Kurtenbach (✉) · I. Prause · C. Weigel · B. Corves
Department of Mechanism Theory and Dynamics of Machines (IGM),
RWTH Aachen University, Aachen, Germany
e-mail: kurtenbach@igm.rwth-aachen.de

I. Prause
e-mail: prause@igm.rwth-aachen.de

C. Weigel
e-mail: weigel@igm.rwth-aachen.de

B. Corves
e-mail: corves@igm.rwth-aachen.de

J. C. García-Prada and C. Castejón (eds.), *New Trends in Educational Activity in the Field of Mechanism and Machine Theory*, Mechanisms and Machine Science 19,
DOI: 10.1007/978-3-319-01836-2_21,

kind of geometry software can be efficiently applied to a present problem where a mechanism has to be developed.

Keywords Mechanisms analysis · Geometry software · Interactive animation

1 Introduction

Dynamic geometry software offers many possibilities for the analysis of mechanisms. It is an alternative to analytical and numerical calculation procedures. Also, it opens up opportunities that are non-existent for mechanism design on paper which tends to be cumbersome, static, often inaccurate and confusing. The fundamentals of different geometry software are various and accordingly diverse is the realized functionality. All programs have in common, that they require a high degree of abstraction for mechanism analysis and synthesis. Summing up the papers' goal is the analysis and evaluation of geometry software for the analysis and synthesis of mechanisms. Recommendations for various types of mechanisms are to be found. It is of interest which programs are fail-safe and if there are programs allowing the user to correct a mistake that was made in the beginning later on.

2 State of the Art

Most dynamic geometry programs offer an interactive model actualization, with which the changes made to a design can be analyzed. This way, the introduced programs can already be of great help during the early phases of mechanism development, especially during the conceptual design. Figure 1 gives an overview on dynamic geometry software. This systematization is divided into groups containing the different programs, which are currently available.

Programs which can be used to analyze planar mechanisms have different target groups and offer various possibilities of conceptual design.

Group A—School geometry software: A rather big number of dynamic geometry software available is developed for the application in school mathematics. Their goal is originally to offer the students a different access to geometry than the simple use of paper, circle and ruler allows. The different programs make geometry ready to be experienced dynamically on different levels.

Group A1—Simple school geometry software: Such software gives the opportunity to create interactive and dynamic circle and ruler constructions and subsequently analyze them. Actively and interactively constructing and manipulating geometric figures (e.g. by using transformations or various variations and dynamic changes) allows the user to visualize and solve geometric problems. Also, an intuitive user interface is important for these programs.

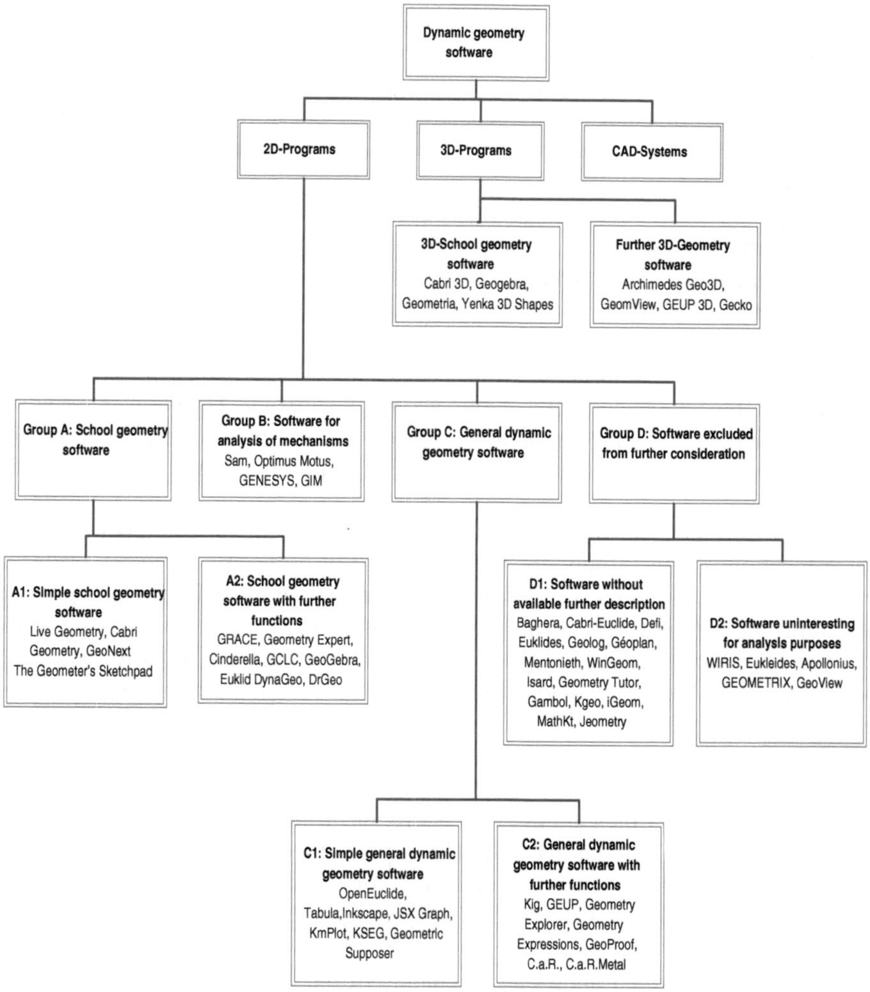

Fig. 1 Overview on dynamic geometry software

Group A2—School geometry software with further functions: Such software offers functions exceeding circle and ruler constructions. These further functions are various and do not allow dividing the programs into further sub-groups. Various geometric constructions or drawings and a following interactive manipulation, respecting the restrictions made during the construction, can be performed. This possibility of working interactively enables the user to perform experiments with the principles of school geometry. The use of colors is supported by most of the programs and helps making especially complex constructions less confusing.

Group B—Software for analysis of mechanisms: Such software is invented for mechanism design and offers functions accordingly. Apart from points, lines, circles and so on, complete gear wheels or partial linkages can be constructed well-directed. In comparison with other groups, the effort for the design of e.g. a crank-and-rocker mechanism is much smaller, since the links can be quickly clicked together without performing rather time consuming geometric circle and ruler constructions. In relation to classical dynamic geometry software there are no geometry tools but links or wheel tools instead.

Group C—General dynamic geometry software: This group summarizes all programs which—in contrast to school geometry software (Group A)—do not have a specific target group reasoning its development. Nevertheless, similarities to the programs of group A can be found. All programs are based on the principle of dynamically adaptable geometry. This group can be divided further into simple geometry software and such with further functions.

Group C1—Simple general dynamic geometry software: The software grouped in this section shows some similarities to group A1 and offers basic functions for generating geometric constructions, which are actually simpler and more basic than in other groups.

Group C2—General dynamic geometry software with further functions: The basic functions of this group are similar to those of group A2. Objects' attributes such as color, style, width or visibility/transparency can be changed and optimize the clarity of a design. The zoom function enables the user to have a closer look at important details, while unlimited undo and redo help the program to gain a better usability. Also, the use of macros can simplify complex design procedures immensely. The possibility of visually modifying constructions directly on screen offers the intended interactivity.

Group D—Software excluded from further consideration: In this group, dynamic geometry programs, which are not interesting for further consideration for the purposes of this paper, are summarized. The reasons for this lie on the one hand in the lack of available information (group D1), on the other hand in the functional orientation (group D2), which makes the programs unfit for the use of mechanism analysis.

The following Table 1 shows a comparison of the programs from the different groups chosen for further analysis. A "+" stands for yes, a "−" for no and a "0" for principle existence/suitability, while n.a. means that the information on this point is not available.

Cinderella as a reference program (Fig. 2 shows an exemplary screenshot of the user interface) is used for comparison, because this software is actually the most widely used software in teaching at IGM of RWTH Aachen University [7–9]. Therefore, it is of interest, if other programs are more suitable to fulfill specific tasks for the design or analysis of mechanisms.

Table 1 Comparison of the selection of analyzed programs

Program	Drag linkage-function	Measurements	Coupler curves/ envelops	Animations	Incl. source code	Exports
A1: GEONExT [1]	+	+	+	+	n.a.	n.a.
A2: GeoGebra [2]	+	+	+	+	+	Web
B: SAM [3]	+	+	+	+	n.a.	n.a.
C1:OpenEuclide [4]	–	–	–	–	n.a.	n.a.
C2: C.a.R. [5]	+	–	+	+	+	Web
Cinderella (A2) [6]	+	+	+	+	0	Web

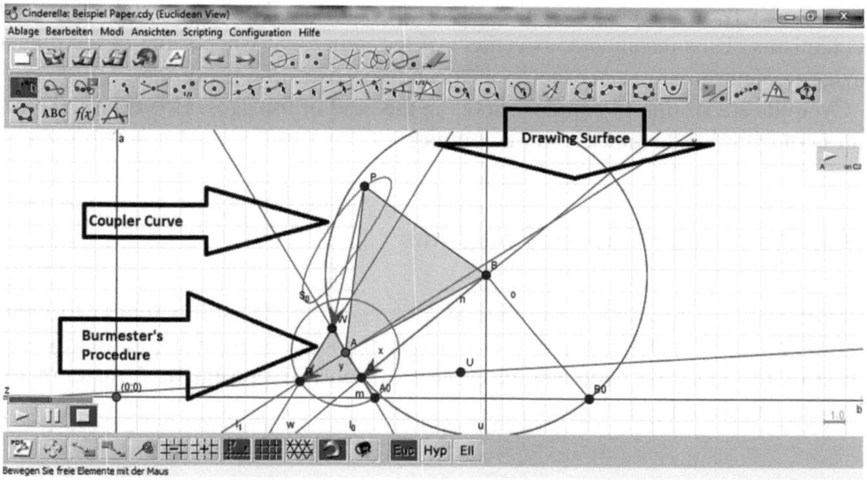

Fig. 2 Screenshot of an example in Cinderella

3 Description of Relevant Analysis Procedures

There are three relevant types of mechanisms, namely linkages, geared mechanisms and cam mechanism. In the following, some analysis procedures for linkages and geared mechanisms using dynamic geometry software are introduced and explained shortly.

The first type of mechanism to be dealt with is the linkage mechanism, which already requires interesting analysis procedures. During the position analysis of the linkage the coordinates of certain joints are to be determined. Additionally, the coupler curve of an arbitrarily given point on the coupler link is to be determined, which is a procedure that would be very cumbersome to do by hand.

The center of curvatures for a given point on the coupler curve can be found using the procedure of BOBILLIER, for example. Velocities and acceleration can be analyzed using specific geometric procedures, such as the Euler procedure, the procedure of rotated velocities, BURMESTER's method or the Theta-line-procedure. One example is BURMESTER's method, stating that the tips of the velocity vectors of three given points on a link form a figure that is similar to the figure formed by the points on the link themselves. Graphical methods for determining accelerations are not analyzed explicitly, because the feasibility of these methods results directly from the feasibility of the methods presented for velocities.

An important characteristic of a mechanism is its transmission ratio. It can be found graphically by using relative instantaneous centers and applying ARONHOLD-KENNEDY-theorem. Setting the appropriate distances of the instantaneous centers into relation, the transmission ratio can be calculated.

Additionally, analysis methods for geared mechanisms will be explained. The first one is the KUTZBACH-scheme, which allows determining the velocity state of geared mechanisms. Thus instantaneous centers of the individual wheels and of the planet carrier are determined, using them to execute the Theta-line-procedure.

Cycloids are similar to coupler curves in the field of linkages. However, the realization of cycloids on paper is more complex.

4 Analysis of the Chosen Programs

In this chapter the programs are examined, looking at their feasibility for the different analysis procedures described in the previous chapter. As an example, the examination of the dynamic geometry program GeoGebra is presented. An exemplary screenshot is shown in Fig. 3.

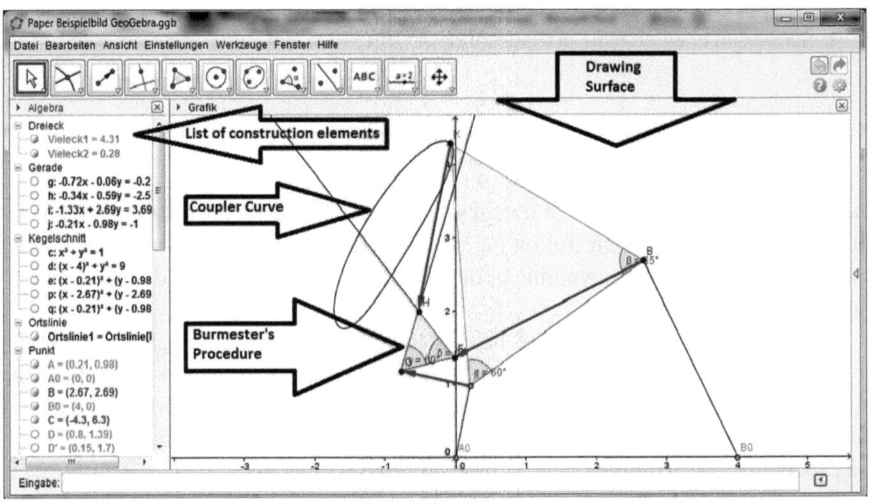

Fig. 3 Screenshot of an example in GeoGebra

The program offers an interface in which the right part of the screen shows the user interface for designing and animating the linkage, while on the left side the geometric construction elements are enumerated including their coordinates. Since the coordinates of every arbitrary point are shown automatically on the left side of the screen, the position analysis is easily feasible.

Coordinates are not shown directly anywhere in Cinderella. Additional orthogonal lines have to be created for the position analysis together with using additional measuring tools, making the procedure more time consuming and cumbersome than in GeoGebra.

GeoGebra's toolbar contains the tool "locus". This tool is used to automatically calculate and create coupler curves. A similar tool exists in Cinderella, as well.

Centers of curvature cannot be calculated automatically in GeoGebra. Their graphical construction using the procedure of BOBILLIER is, however, easy and quick. Construction elements, which are no longer required to be visible, can easily be made invisible directly by appropriate selection from the list on the left side of the screen, thus enhancing appearance and clarity of the mechanism analysis.

In comparison, especially the construction and reconstruction of angles takes more time in Cinderella than it does in GeoGebra. Also, the lack of a mouse-integrated zoom and pan function steals time. Turning objects invisible also takes more time.

GeoGebra does not offer the possibility to automatically calculate velocities, but the geometric analysis of velocities, using velocity vectors, which are shown with an arrowhead—on the left side of the screen split into the two coordinate directions—is possible. The aforementioned procedures are feasible all the same.

There is no function for automatic calculation of transmission ratio in GeoGebra. The construction of relative instantaneous centers does not cause any problems. Only the clarity of the construction could be optimized concerning the appellation. In Cinderella the construction also works quite easy, as it does in GeoGebra. The difference that makes GeoGebra more applicable for this procedure is the mouse-integrated zoom and pan function. It makes bigger constructions easier to oversee and more comfortable to work with.

Distances of instantaneous centers can easily be measured. The measuring result is shown directly in the user interface for designing the mechanism. The results can be exported into a table within GeoGebra. This works quite similar in Cinderella, the measuring results are also shown in the user interface.

The construction of the KUTZBACH-scheme for velocities, including the construction of the gear mechanism model works very well. In comparison, the construction is more time-consuming when using Cinderella. Also, problems may occur if the construction is not done in a special order, since picking out special lines can become problematic.

The generation of cycloids works rather easy in GeoGebra. First, an appropriate mechanism has to be designed. Then, specific points are given their specific velocities. This is done in the object properties where the joints to be animated can be selected and the different joint velocities are defined. Finally, it can be chosen if a joint's locus shall be displayed. For wheel mechanisms this results in a cycloid after starting the

Table 2 Comparison of dynamic geometry programs

Program	Position analysis	Coupler curves/center of curvatures	Velocities/ accelerations	Transmission ratio/relative instantaneous center	Procedure of Kutzbach/ velocities	Cycloids
A1: GEONExT	+	0/0	0	0	0	+
A2: GeoGebra	+	+/+	+	0	+	+
B: SAM	+	+/+	+	0	+	+
C1: OpenEuclide	0	−/−	−	−	−	−
C2: C.a.R.	+	+/+	0	−	+	0
Cinderella	0	+/0	0	0	0	+

animation. The procedure works almost the same way in Cinderella, only the display of the cycloid has a different design.

5 Conclusions

At this point a comparison of the programs respectively the different groups is done. Table 2 visualizes the analysis procedures feasible with the sample programs of the different groups. "+" denotes a very good adequacy, "0" means that the procedure is suitable in principle, but with a more complicated implementation and "−" denotes no or very bad suitability. It shows that the program OpenEuclide representing the group C1 is absolutely unsuitable for any mechanism theory analysis purposes, since basic geometric tools which would be necessary for analysis purposes are missing. Therefore, it will not be commented further.

The position analysis works very well and easy in the groups A1, A2, B and C2. Overall, the groups A2 and B are the ones with the best results. They show a similar weakness in the determination of transmission ratios for crank mechanisms. In particular this does not mean that it is impossible, but the approach could be optimized.

Group A1 can be used for almost all purposes, but goes along with a bigger outlay. Overall it is suitable for mechanism analysis, but is not the best of all choices. Group C2 is very well suitable for a few procedures and unsuitable for others. For this group it is important to take the exact task at hand into account before evaluating the applicability for mechanism analysis.

The cycloids stand out as well as the position analysis. For the first one, almost all programs are suitable rather well, with the exception of Cinderella. Regarding the cycloids, applying C.a.R. the user has to face some difficulties, especially in relation to the other programs, generating cycloids quite easy.

The overall winners are the groups A2 and B, followed by the group C2. In some cases, the group C2 even offers better clarity and is sometimes more intuitive.

References

1. http://geonext.uni-bayreuth.de/index.php?id=2394 (2013). Accessed 14 Mar 2013
2. http://www.geogebra.org/cms/de/info (2013). Accessed 14 Mar 2013
3. http://www.artas.nl/de/sam/sam-fur-spezialisten (2013). Accessed 14 Mar 2013
4. http://coulon.publi.free.fr/openeuclide/ (2013). Accessed 14 Mar 2013
5. http://zul.rene-grothmann.de/doc_de/Informationen.html (2013). Accessed 15 Mar 2013
6. http://www.cinderella.de/tiki-index.php (2013). Accessed 14 Mar 2013
7. Corves, B.: Computer-aided lectures and exercises: graphical analysis and synthesis in mechanism theory. In: Proceedings of the 11th World Congress in Mechanism and Machine Science, vol. 2. China Machine Press, Tianjin, China, 1–4 April 2004
8. Corves, B., Riedel, M., Hüsing, M.: Descriptive and intuitive mechanism design and synthesis using geometry-based computer-aided methods. In: 3rd World Congress in Mechanism and Machine Science, Guanajuato, México: International Federation for the Promotion of Mechanism and Machine Science, 19–25 June 2011
9. Richter-Gebert, J., Kortenkamp, K.H.: Benutzerhandbuch für die interaktive Geometrie-Software Cinderella Version 1.2. Springer, Berlin (2000)

References

[illegible]

Kinematic Analysis of Planar Mechanisms by Means of Examples

M. Urízar, O. Altuzarra, M. Diez, F. J. Campa and E. Macho

Abstract The aim of this work is to approach the difficulties students usually encounter when facing up to kinematic analysis of mechanisms. A deep understanding of the kinematic analysis is necessary to go a step further into design and synthesis of mechanisms. We can conclude from experience that supporting and complementing the theoretical lectures with specific software is really helpful. In this sense, software is used during the practical exercises, serving as an educational complementary tool reinforcing the knowledge acquired by the students. Several questions are outlined to the students, so that they are encouraged to justify the validity of their results. GIM software performs kinematic analysis and motion simulation of planar mechanisms. The main capacities of the software are: solving the position problem, computing velocities and accelerations, singular analysis, and visualization of instantaneous center of rotation, acceleration pole, curvature center and circle, fixed and moving centrodes and main circles. The graphical representation of all results favors the learning of the theoretical concepts explained in the subject and also, stimulates the critical reasoning the students must acquire.

Keywords Education · Mechanism and machine theory · Kinematic analysis · Motion simulation

1 Introduction

Mechanism and Machine Theory plays an essential role in education on Mechanical Engineering. In the Faculty of Engineering in Bilbao, University of the Basque Country (UPV/EHU), two main subjects can be highlighted in this field, which are Applied

M. Urízar (✉) · O. Altuzarra · M. Diez · F. J. Campa · E. Macho
Faculty of Engineering in Bilbao, Department of Mechanical Engineering,
University of the Basque Country (UPV/EHU), Bilbao, Spain
e-mail: monica.urizar@ehu.es

J. C. García-Prada and C. Castejón (eds.), *New Trends in Educational Activity in the Field of Mechanism and Machine Theory*, Mechanisms and Machine Science 19,
DOI: 10.1007/978-3-319-01836-2_22, © Springer International Publishing Switzerland 2014

Mechanics (Statics, Kinematics and Dynamics) [1, 2], and Kinematics of Mechanisms [3]. The subject of *Applied Mechanics*, taught during the second course in Mechanical Engineering, covers the following significant topics: center of gravity and related properties, moments of inertia, statics, kinematics and dynamics of rigid body, velocity and acceleration fields, and planar motion analysis. *Kinematics of Mechanisms,* which is included in the third course, teaches all the necessary notions of kinematics for planar and spatial mechanisms (planar motion geometry, graphical constructions, analytic methods, transformation matrices for spatial mechanisms, etc.), as well as dimensional synthesis of planar mechanisms. In general, students encounter some difficulties when facing up to kinematic analysis of mechanisms, due to the complex mathematical and geometrical concepts intrinsic to the theoretical bases. From experience, it can be concluded that computer programs help students to achieve a better understanding of the theoretical notions explained in the lectures [4, 5]. Indeed, currently, most of the textbooks of Mechanism and Machine Theory include simulation programs to reinforce and complement the contents of the book [6, 7].

GIM is a registered software for the kinematic analysis of planar mechanisms created by the COMPMECH Research Group belonging to the Department of Mechanical Engineering of UPV/EHU. The software is intended for educational purposes, in particular destined to the field of kinematic analysis and motion simulation of planar mechanisms. Mechanisms with *n*-ary elements joined by revolute and prismatic pairs can be introduced. The position problem is solved iteratively using a numerical method, several of its conditions can be controlled and visualized. Inputs can be introduced with a polynomial up to the quintic, so position, velocity and acceleration can be specified at motion extremes. Paths of points and the area swept by an element can be plotted. Velocities and accelerations can be obtained and depicted as the motion is performed, also graphs and tables with this data can be plotted. Every kinematic construction can be represented, instantaneous centre of rotation, fixed and moving polodes, acceleration pole, acceleration circles, and so on.

2 GIM

GIM software has two main modules: *Geometry* and *Motion*. *Geometry* module is the one in charge of defining a specific design of the mechanism object of study. The other module, *Motion* module, performs the kinematic analysis and motion simulation of the mechanism.

In Fig. 1, the starting main window associated with *Geometry* module is shown, highlighting in a close-up view the options inside this module. So as to define a desired design the user selects among the following icons and plots on the main screen: establishing the nodes of the mechanism, creating the elements, **Au** automatic assignment of joints, defining revolute joints and prismatic joints, establishing prismatic limbs, creating a fixed joint, modifying location of points, and **X** delete.

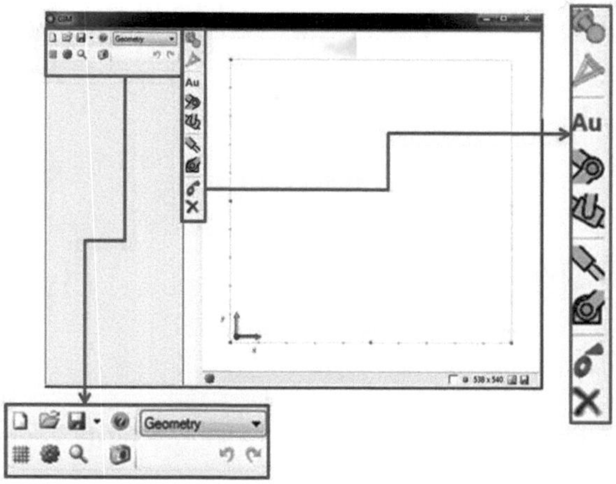

Fig. 1 Main window of GIM software. Geometry module

Once the mechanism's geometry has been established, *Motion* module will perform the kinematic analysis of the mechanism. In Fig. 2, *Motion* module main screen, using the four-bar mechanism as an illustrative example, is shown. The motion of the mechanism according to a specific input can be visualized by the user. In the example under analysis, the fixed revolute joint on the left is chosen as the actuated one, as highlighted in Fig. 2. The user can define the input conditions (position, velocity and acceleration), appearing the corresponding polynomial input on the right-hand side (see Fig. 2).

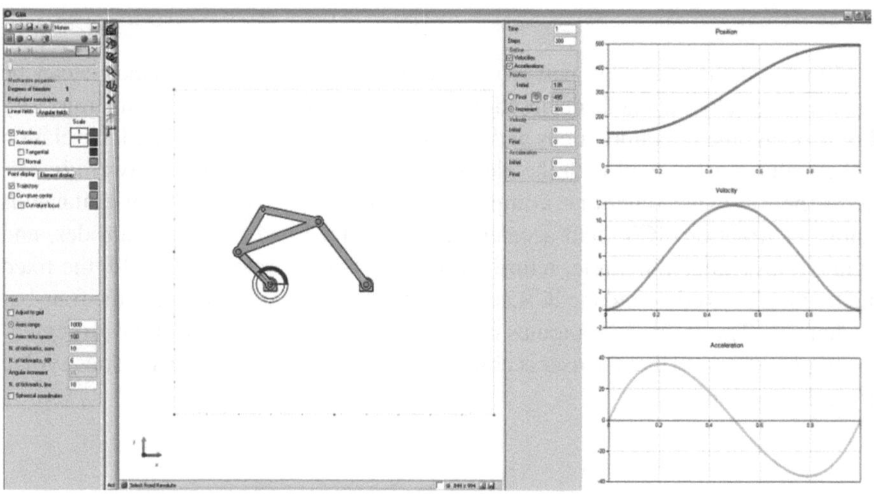

Fig. 2 Motion module. Four-bar mechanism example

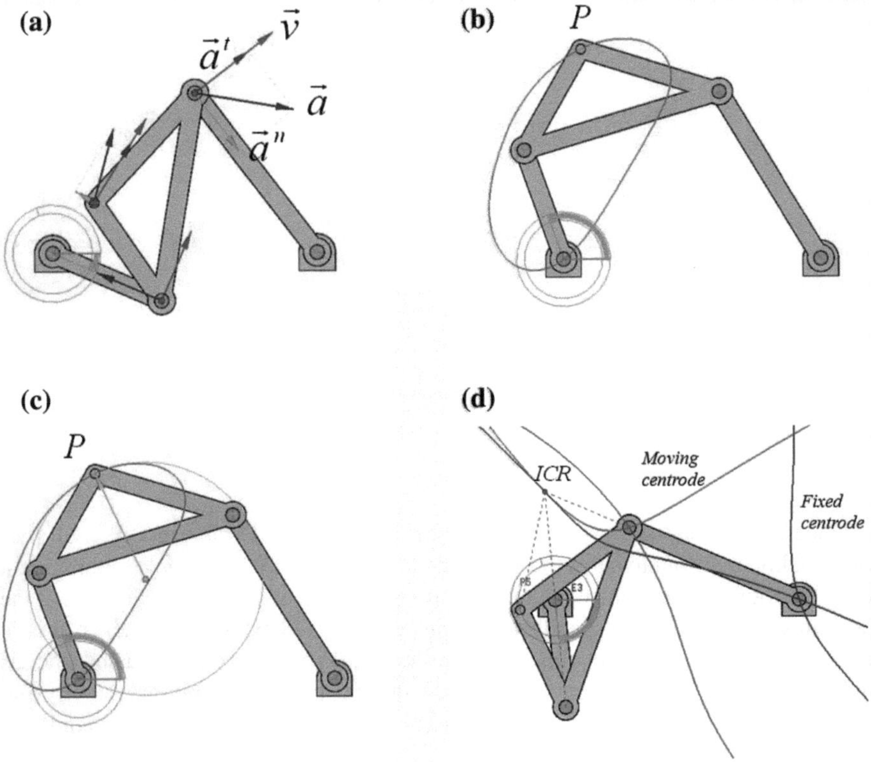

Fig. 3 Several options displayed in Motion module

Significant kinematic properties can be analyzed. On the one hand, it can be obtained the velocities and accelerations of points of interest (Fig. 3a), its trajectory (Fig. 3b), and the curvature locus and center (Fig. 3c). On the other hand, in relation to the features associated with the elements of the mechanism, the software implements the following functions: computation and visualization of the Instantaneous Center of Rotation (ICR) and acceleration pole, fixed and moving centrodes, and main circles (inflection circle, return circle and Bresse circle). In Fig. 3d, the fixed and moving centrodes, and the ICR, are depicted for a certain position of the four-bar mechanism. The software computes the continuous tracking of any of the aforementioned properties, so that the user is able to visualize how these characteristics evolve along the motion.

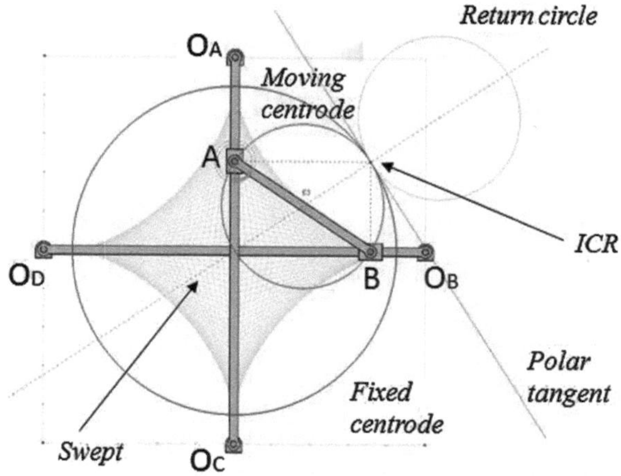

Fig. 4 Illustrative example: PRRP mechanism

3 Application

The theoretical lectures are complemented and reinforced with practical exercises in which GIM software is used. The target is to help the students understanding all the concepts illustrated in the lectures.

The practical exercises are organized as follows. Initially, the operation of the software is explained step-by-step to the students. To do so, the planar PRRP mechanism represented in Fig. 4 is chosen as an illustrative example.

A video showing the capacities of GIM software, making use of the aforementioned mechanism is loaded in the web site of COMPMECH Research Group: http://www.ehu.es/compmech/software/.

In a second stage of the exercise, the students must handle the software designing and performing the kinematic analysis of a specific planar mechanism. In this part, the students are encouraged to try all the possibilities the software offers.

Finally, a document including several different questions related to the subject is given to the students. Some designs of planar mechanisms, as the ones shown in Fig. 5, are proposed. Students must test and justify the validity of the obtained results. The following questions are outlined:

- Analyze the four-bar mechanism (Fig. 5a) regarding Grashof criterion. On the one hand, define the geometry parameters so that the mechanism satisfies Grashof criterion. Specify the relation among the dimensions of the bars so that $\mathbf{O_A A}$ can rotate a full turn and $\mathbf{O_B B}$ has a pendulum motion. For a generic posture, obtain the moving and fixed centrodes, ICR, and main circles.

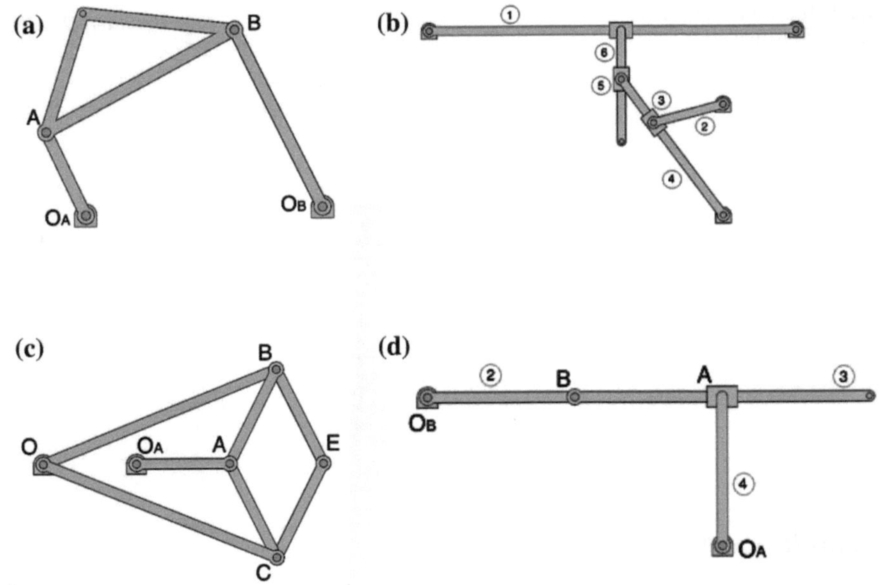

Fig. 5 Proposed planar mechanisms during the practical exercise

- On the other hand, design a non-Grashof four-bar mechanism. Obtain all the locking postures, and define the sequence of inputs that must be given so as to unlock the mechanism.
- For the mechanism represented in Fig. 5b, named the *Scotch yoke*, represent the graphical plots of velocities and accelerations of the different elements, justifying their linear or angular units.
- Design the *Peaucellier* mechanism, shown in Fig. 5c, in which the following relations must be fulfilled: $\mathbf{OB} = \mathbf{OC}$, $\mathbf{AB} = \mathbf{BE} = \mathbf{CE} = \mathbf{AC}$ and $\mathbf{OO_A} = \mathbf{OA_A}$. Represent the trajectory of point E, verifying it is vertical. For the initial pose (Fig. 5c), obtain and represent the inflection circle, return circle and Bresse circle. Visualize the variation of these main circles along the motion of the mechanism.
- For the planar mechanism depicted in Fig. 5d, obtain all velocity poles. Depict the main circles associated with the third element, for the pose indicated in the figure. Represent the evolution of the main circles for a full cycle of the mechanism. Justify if any of these circles degenerates into a line for any specific pose of the mechanism.

In Fig. 6, the results according to some of the explained questions are shown.

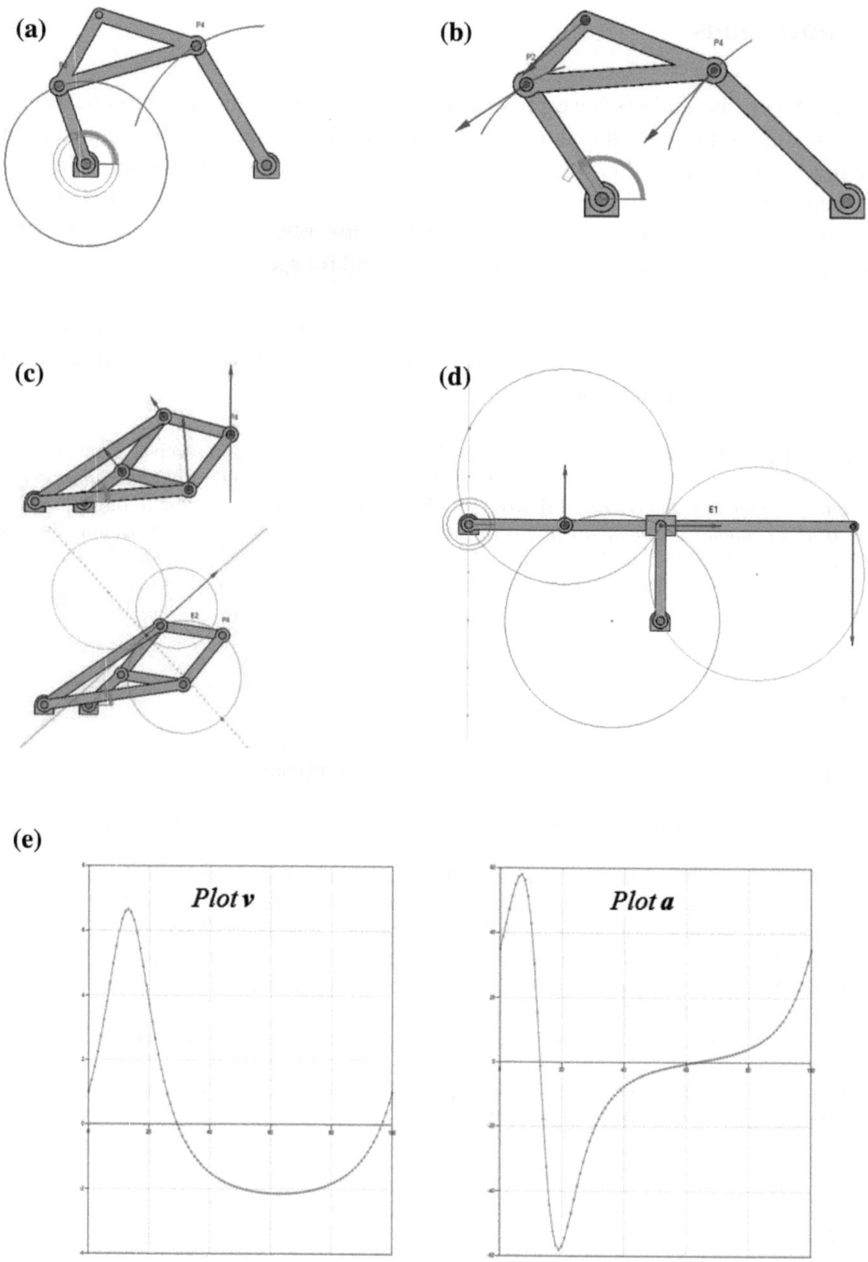

Fig. 6 Visualization of some results of the practical exercises

4 Conclusions

In this paper, the authors have presented a software tool intended for educational purposes in the field of kinematics of mechanisms, GIM software. Implemented in the practical exercises, this software serves as a complementary tool reinforcing the theoretical concepts explained in the lectures. Handling the software the students are able to analyze several different planar mechanisms, obtaining and visualizing the main graphical constructions along their motion and for specific postures, computing the velocities and accelerations of points of interest, determining the trajectory of points of interest, and so on. GIM software has proven to be a very effective tool that attracts the attention of the students, and also makes easier for them to understand all the notions taught in the lectures.

Acknowledgments The authors wish to acknowledge the financial support received from the University of the Basque Country (UPV/EHU) through the *Vicerrectorado de Calidad e Innovación Docente y Competitividad* (Project PIE2012-14/ 6564), the Spanish Government through the *Ministerio de Economía y Competitividad* (Project DPI2011-22955), the European Union (Project FP7-CIP-ICT-PSP-2009-3) and the Regional Government of the Basque Country through the *Dpto. Educ., Univ. e Investig.* (Project IT445-10) and UPV/EHU under program UFI 11/29.

References

1. Bilbao, A., Amezua, E.: Mecánica Aplicada: Estática y Cinemática. Editorial Síntesis, Madrid (2008)
2. Bilbao, A., Amezua, E., Altuzarra, O.: Mecánica Aplicada: Dinámica. Editorial Síntesis, Madrid (2008)
3. Hernández, A.: Cinemática de mecanismos: Análisis y Diseño. Editorial Síntesis, Madrid (2010)
4. Pennestrì, E. and Vita, L.: Advances in Computational Multibody Systems, Chapter: Multibody dynamics in advanced education, pp. 345–370. Springer, Netherlands (2005)
5. Petuya, V., Macho, E., Altuzarra, O., Pinto, Ch. and Hernández, A.: Educational software tools for the kinematic analysis of mechanisms—computer applications in engineering education (2011). doi:10.1002/cae.20532
6. Waldron, K.J., Kinzel, G.L.: Kinematics, dynamics and design of machinery. Wiley, New York (1999)
7. Norton, R.L.: Design of machinery: an introduction to the synthesis and analysis of mechanisms and machines. McGraw-Hill, Boston (2003)

Teaching Mechanism and Machine Theory with GeoGebra

X. Iriarte, J. Aginaga and J. Ros

Abstract Teaching Mechanism and Machine Theory involves the description of mechanisms which are not easy to represent with static drawings. Computer applications are a common way to reproduce the motion of mechanisms. GeoGebra is a free dynamic geometry software with interactive graphics that can help in this task. By introducing mathematical constraints, such as constant distances, intersections, tangency or perpendicularity, it is possible to build a wide range of interactive mechanisms. This software has been used at the Public University of Navarra on a Degree in Mechanical Engineering, having a successful experience with the students. The software has been used in two ways. On the one hand, some mechanisms have been prepared and shown to the students in theoretical lessons in order to explain their motion. On the other hand, practical lessons have been carried out, in which the students have programmed one degree of freedom mechanisms with GeoGebra.Students have shown interest in the practical lessons and results have been satisfactory.

Keywords Mechanism and machine theory · GeoGebra · Dynamic geometry · Interactive mechanisms · Practical lessons

1 Introduction

In the subject of Mechanism and Machine Theory (MMT) of a common Mechanical Engineering Degree, most of the topics that are taught to the students are related to the relative movement of different parts of a mechanism or machine. Analysis and

X. Iriarte (✉) · J. Aginaga · J. Ros
Public University of Navarra,Pamplona, Spain
e-mail: xabier.iriarte@unavarra.es

J. Aginaga
e-mail: jokin.aginaga@unavarra.es

J. Ros
e-mail: jros@unavarra.es

J. C. García-Prada and C. Castejón (eds.), *New Trends in Educational Activity in the Field of Mechanism and Machine Theory*, Mechanisms and Machine Science 19,
DOI: 10.1007/978-3-319-01836-2_23,

synthesis of linkages, cams and gears are customary topics in these subjects, and geometric methods have been developed for decades and are explained in most of the classic MMT books [7, 8]. However, all the concepts related to movement are explained in books with static figures, and the movement is represented with arrows, shades or multiple delayed drawings of the same system. These representations are usually clear enough for experienced engineers and researchers, but happen to be insufficient for students with low visualisation skills, making them difficult to understand the underlying concepts. The different interpretation that teachers and students do of a figure is referred to as *Illusion of Transparency* and has been analysed by several educational researchers [5]. Modern multimedia tools have shown to be very useful to overcome these difficulties and their use is increasing in all teaching environments. Videos, animations and simulations can show much clearer the concepts that are to be explained.

One of the most promising tools to overcome movement constraints in MMT lessons are the Dynamic Geometry Environments (DGE) that allow to build geometric entities in terms of some variables, and afterwards rebuild automatically the whole geometric construction by changing the values of the variables. GeoGebra [3] is one the most trending DGE-s used in teaching environments and the one used in this study.

Despite the outstanding potential of the DGE-s to teach MMT, almost no references or material have been systematically developed in this field [1, 6]. In this paper, the utility of DGE-s as GeoGebra is shown for teaching MMT. Section 2 deals with the characteristics of GeoGebra as a DGE for teaching MMT. Section 3 shows different applications of the DGE-s related to MMT, and in Sect. 4 we show how we have used GeoGebra to teach MMT in both the theoretical and practical lessons in the Mechanical Engineering Degree at the Public University of Navarra.

2 GeoGebra as an Instrument for Teaching MMT

GeoGebra can be defined as an interactive geometry software with algebra and spreadsheet capabilities. It includes most of the main construction tools of a 2D CAD software but with the key advantage of being able to build the so called *Dynamic Models*, i.e. geometric constructions based on certain parameters that can subsequently change their value.

The *Dynamic Models* are customary in MMT. All the mechanisms (linkages, cams, gears ...) can be represented as *Dynamic Models* in which the parameters play the role of coordinates. Continuously varying these coordinates, the movement of the mechanism is represented. Additionally, there are many mechanism analysis procedures based on geometry that accept or discard a certain design in terms of the values of some parameters that play the role of design variables. Using a trial and error approach, the influence of some of the design variables on the desired performance can be quickly analysed in an interactive way.

The advantages of GeoGebra as a tool for teaching MMT are many. It runs on Windows, Linux, and Mac operative systems and it also runs on line (directly in a web browser). Moreover, being free software, any student can install and use it free of charge. But, the benefits of free software are not usually constrained to the software itself. The GeoGebra community is a wide group of people (including software developers, artist, teachers and enthusiasts) that have created forums (with thousands of posts) events and conferences, and even regional and national GeoGebra institutes where the use of GeoGebra is promoted. Furthermore, the users can access the GeoGebratube web page [2], which is the official repository of GeoGebra constructions and GeoGebra related resources with more than 25,000 materials. Last but not least, the software and the documentation are available in many different languages. The 2D constraints of GeoGebra are being overcomed by GeoGebra3D. However, the first stable version is still to be released.

GeoGebra has been mainly used in primary and secondary schools, but since 2008 it has been also used in the Public University of Navarra in the Masters for Primary School Teachers Education [4]. In this paper its use at university level has been extended to the Engineering Degree for teaching MMT.

3 DGE-s Applications for Mechanism and Machine Theory

The DGE-s can help teaching many of the concepts included in a customary MMT subject. Starting from the very basic concepts regarding a mechanism, and ending with the geometric constructions to synthesise a linkage. In this section we will show some of the GeoGebra applets that have been developed to teach MMT in the Public University of Navarra.

3.1 Kinematics of Linkages

The basic kinematic concepts of linkages can be easily introduced with a simple four bar linkage, moving the crank by dragging a point attached to it. Customary concepts such as kinematic inversions, Grashoff's Law, mechanical advantage and couplers curve, can be easily visualised by similar constructions. Figure 1 shows a four bar mechanism together with some sliders that change the lengths of their links. Grashoff's Law can easily be checked observing the mobility for different values of the lengths.

Similarly, Fig. 2 shows a four bar mechanism with a P point fixed to the coupler. Depending on its location with respect to the couplers segment, point P will describe a different curve of the coupler. When the location of point is decided by dragging P, the mechanism can be moved by dragging point Q.

Some of the classic mechanisms have also been built for the students to understand their kinematics. Applets for the straight line mechanisms of Watt, Roberts,

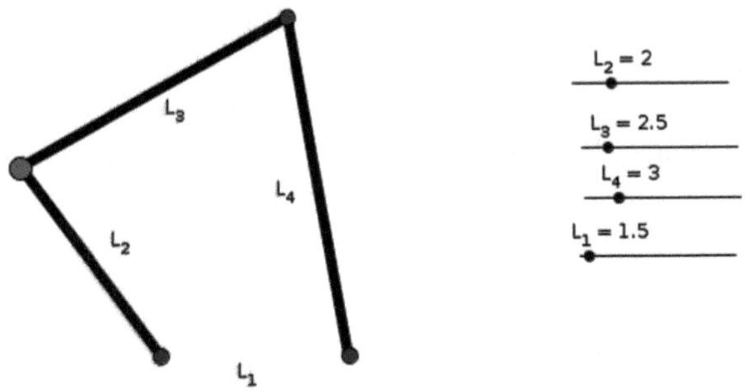

Fig. 1 Four bar mechanism with variable link length to check Grashoff's Law

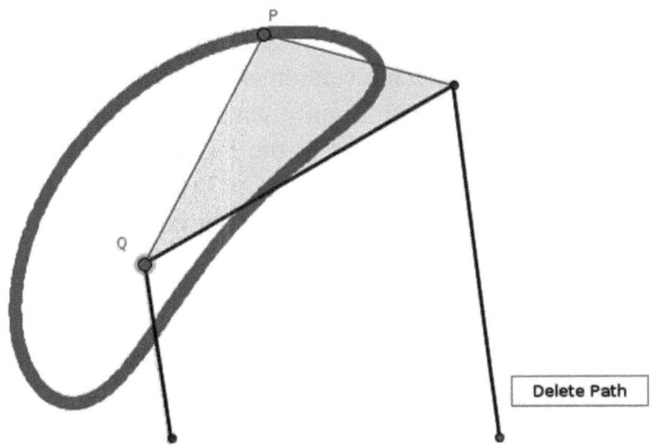

Fig. 2 Curve described by a certain point of the coupler

Chevyshev, Peaucillier, and Hart, and the quick-return mechanism of Whitworth have been built.

Other concepts as singularity configurations and bifurcations can be easily introduced to the students through a DGE. In GeoGebra, when the moving element exceeds a blocking singular configuration, the mechanism can not be assembled, and GeoGebra makes it evident by hiding the links that can not satisfy the geometric constraints. On the contrary, when a bifurcation occurs, the user can not control the path to follow but depending on the evolution of the iterative numerical algorithm, it will sometimes follow one path or the other. Figure 3 shows Hart's mechanism and the different paths that point P has described while moving point A. For the configuration in which point A is in the line O_1O_2, all the links are aligned to each other, and the bifurcation allows P to follow three different paths in each direction: a curve towards the right, a curve towards the left, and a vertical straight line.

Some velocity concepts can also be described by geometric constructions. The location of the Instantaneous Centre of Rotation (ICR) for instance, can be geometrically calculated drawing the lines that pass through the poles of different couples of links. Moreover, the movement of the whole geometric construction (mechanism and lines for the calculation of the ICR-s) renders it evident that the ICR-s depend

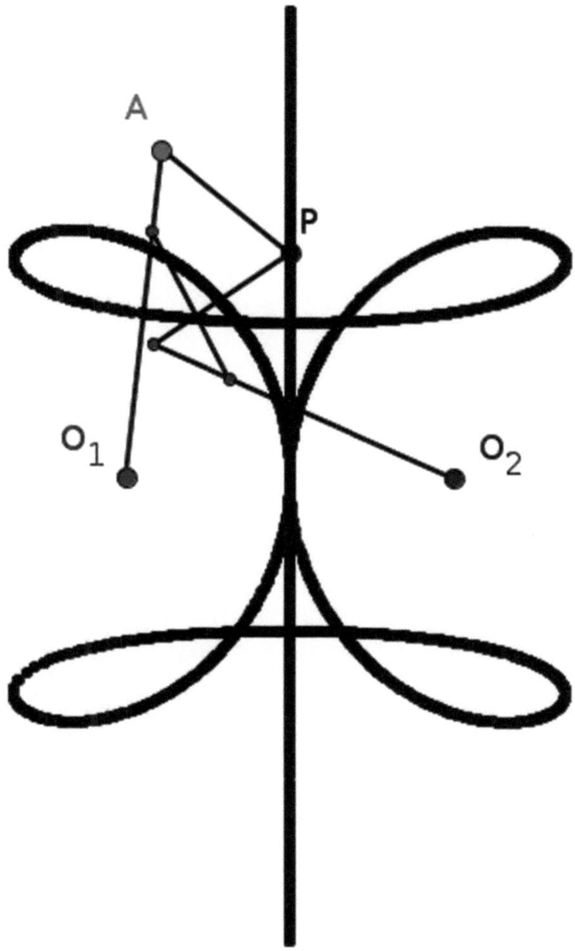

Fig. 3 Paths of point P for Hart's mechanism

on the pose of the mechanism. Figure 4 shows an applet with the poles and ICR-s of a four bar mechanism.

The analysis of a disc rolling on a surface without sliding also has an underlying velocity concept to be understood. The rolling without sliding constraint is non-holonomic. Therefore, it has to be defined at velocity level, imposing one of the following constraints: $\boldsymbol{V}_{Ground}(P_{Disc}) = \boldsymbol{V}_{Ground}(P_{Ground})$ or $\boldsymbol{V}_{Disc}(P_{Geom}) = \boldsymbol{V}_{Ground}(P_{Geom})$. Therefore, it has to be explained to the students that P_{Disc} is the point that kinematically belongs to the *Disc* and that P_{Geom} is the geometric contact point between the *Disc* and the *Ground*. These concepts render evident when the points are independently drawn for an arbitrary position, and coincide in the same point of the plane when the disc is moved towards the analysed configuration. Figure 5 shows the applet used to teach these concepts in a straight forward way.

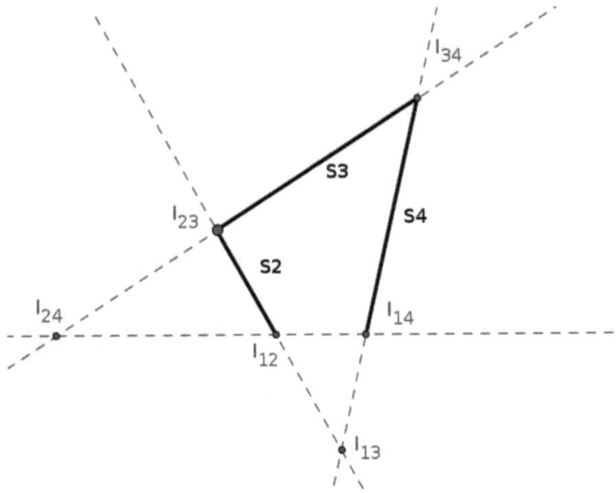

Fig. 4 The poles and ICR-s of a four bar mechanism

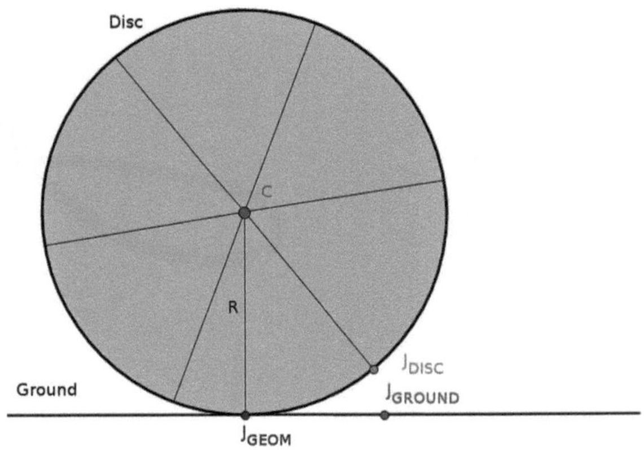

Fig. 5 A disc rolling without sliding in a plane

3.2 Synthesis of Mechanisms

For the synthesis of mechanisms, many geometric procedures are available in the classic MMT books [7, 8]. These geometric procedures are very useful for the students to visually understand the main ideas of the synthesis. Generally speaking, DGE-s can be helpful to solve and understand Function Generation, Path Generation and Body Guidance problems in a geometric context. However, once the geometric construction has been done, it is not always straight forward to understand that even if the procedure holds for any initial data, the obtained mechanism may not satisfy the required constraints.

This is the case for the simple three pose body guidance problem, in which the solution obtained can suffer *branch defect*, i.e. the coupler would reach a certain pose if it would have been assembled in another of the possible branches. Figure 6 shows a three pose body guidance problem resolution in which branch defect occurs. Notice

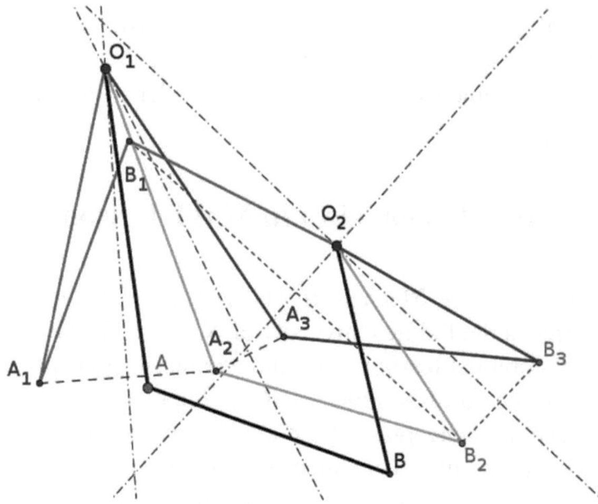

Fig. 6 A three pose body guidance problems resolution with branch defect solution

that the four bar mechanism O_1ABO_2 (in black) will pass through poses two (in green) and three (in blue), but the branch defect will be evident for pose one (in red). The GeoGebra model is a very useful instrument to move the solution-mechanism to check if it passes through the three prescribed poses.

The DGE-s can also be used for the visualisation of dwell or double-dwell mechanisms. These can be synthesised based on the straight or circular curve of a point of the four bars coupler. Building the whole mechanism with a DGE, the curve of the couplers point can be traced in order to show that the generalised dyad remains stopped while the point of the coupler goes over the straight or circular curve segment.

Figure 7 shows a four bar mechanism where couplers point P goes over two nearly straight segments for a complete rotation of the crank. Appropriately adding a

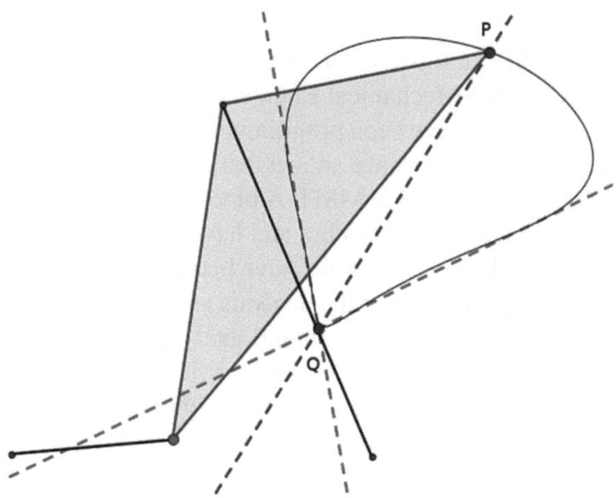

Fig. 7 Synthesis of a double-dwell mechanism adding a rod-slider couple to a four bar

rod-slider couple, a double-dwell mechanism can be built in a straight forward way. The GeoGebra applet renders it evident that when point P goes over the nearly straight segments, the rotation of the slider with respect to point Q remains stopped.

4 Usage of GeoGebra in MMT Lessons

When teaching Mechanism and Machine Theory in the Mechanical Engineering Degree at the Public University of Navarra, GeoGebra has been extensively used. In the theoretical lessons, after the geometric concepts have been taught in the blackboard, the corresponding GeoGebra applet is shown to completely understand the underlying concepts. All the applets are available to the students through the web page of the subject for them to play at will. Moreover, they can not only interactively use the applets, but also download the sources and modify them to build new mechanisms.

Additionally, GeoGebra is used in a 2 h practical lesson. In the first hour, the students are taught how the software works and two simple mechanisms are built (a slider-crank and a four bar mechanism). After that, they are asked to solve two synthesis problems: the three pose body guidance problem, and a dwell mechanism synthesis in which they have to add a dyad.

The students have appreciated the GeoGebra applets and have found it useful for the comprehension of the mechanisms. The practical lessons have also shown that GeoGebra has a steep learning curve (learning is achieved very quickly) and the concepts related to mechanisms and synthesis can easily be taught through it. Most of the students finished the practics in time and understood the concepts.

5 Conclusions and Further Developments

The GeoGebra software has been used as an aid for teaching Mechanism and Machine Theory in the Mechanical Engineering Degree at the Public University of Navarra. Many applets have been programmed regarding kinematics and synthesis of mechanisms. The applets have shown that the software can be very helpful when teaching the concepts related to MMT. A practical lesson has been done in which the students have learnt to use GeoGebra and have also done their own applets solving two synthesis problems. The results have been satisfactory and the students have learnt the underlying concepts of the analysis and synthesis of mechanisms.

Further developments should analyse systematically the effect of using a DGE on the MMT lessons. For instance, it would be very interesting to quantitatively measure the impact that using GeoGebra has on students skills to solve kinematics and synthesis problems.

References

1. Bryant, J., Sangwin, C.: How round is your circle? http://web.mat.bham.ac.uk/c.j.sangwin/howroundcom/front.html (2013)
2. GeoGebra Community: Geogebratube: the official repository of geogebra constructions and geogebra related resources. http://www.geogebratube.org/(2013)
3. Hohenwarter, M.: Geogebra. free mathematics software for learning and teaching. http://www.geogebra.org/cms/(2013)
4. Lasa, A., Saénz de Cabezón, A., Wilhelmi, M.: Restrictions in the lecture of geometry with GeoGebra in Primary School Teacher Education. In: International GeoGebra Conference. University of Upper Austria, Linz (2009)
5. Lasa, A., Wilhelmi, M.: GeoGebra en la formacin de profesorado de E.S.O. y Bachillerato. Cónica 3 (2012)
6. Mora Sanchez, J.: Matemáquinas. La Geometría de los Mecanismos (in spanish). http://jmora7.com/Mecan/index.htm(2013)
7. Norton, R.: Design of Machinery: An Introduction to the Synthesis and Analysis of Mechanisms and Machines. McGraw-Hill, New York (1999)
8. Shigley, J., Uicker, J.: Theory of Machines and Mechanisms. McGraw-Hill Book Company, New York (1981)

Enhancing Mechanism and Machine Science Learning by Creating Virtual Labs with ADAMS

J. L. Suñer Martínez and J. Carballeira

Abstract This work presents a virtual lab created by using the multibody dynamics simulation software ADAMS©, which has been developed to enhance the students experience when learning Mechanism and Machine Science fundamentals. When dealing with spatial kinematics and dynamics, there are some concepts which are usually difficult to visualize for the students. In this way, the use of a simulation software can help to overcome this problem. A set of exercises has been proposed in which the students have to build some models with ADAMS© and then solve some problems related to the theory. This proposal shows to be more attractive for the students, who learn the basics of an engineering tool that is commonly used in the industry and are able to visualize and consolidate the concepts taught at theory classes more easily. But there are also some drawbacks regarding the difficulties that appear when developing these exercises from the teaching point of view. The complexity of the program can complicate the creation of the models in a way that the students can not build them correctly and get the results expected. So, the exercises must be designed properly in order to make them robust enough to avoid these difficulties.

Keywords Learning · Simulation · Virtual labs · ADAMS · Engineering education

1 Introduction

The use of simulation programs is an efficient way to visualize how mechanical systems work and to understand the relation between theory and practice. There several examples of the successful implementation of virtual labs in engineering education,

J. L. Suñer Martínez (✉) · J. Carballeira
CITV, Universitat Politècnica de València, València, Spain
e-mail: josuner@mcm.upv.es

J. Carballeira
e-mail: jacarmo@mcm.upv.es

J. C. García-Prada and C. Castejón (eds.), *New Trends in Educational Activity in the Field of Mechanism and Machine Theory*, Mechanisms and Machine Science 19,
DOI: 10.1007/978-3-319-01836-2_24,

and in mechanical engineering particularly [2, 3]. The analysis and demonstrations that can be performed by using these tools usually exceed the possibilities of physical experiments, improving the students experience and outcomes [1]. Moreover, the effectiveness of this methodology when compared with hands-on laboratories appears to be very similar, where effectiveness refers to the students understanding of the physics of the experiments [4].

The aim of this paper is to describe the methodology used in the lab works of several courses related to Mechanism and Machine Science within the degrees in Mechanical and Aeronautical Engineering offered by the *Universitat Politècnica de València*. In these courses, a set of exercises are developed during the lab hours by using the multibody dynamics simulation software ADAMS©, with the objective of enhancing the students experience when learning Mechanism and Machine Science fundamentals.

First, some exercises designed to improve the understanding of 3D kinematics fundamentals are shown. These exercises will help the students to visualize the motion and to learn how to use the equations of spatial kinematics properly, in order to solve the problems proposed by comparing the numerical results given by the simulations with the analytical solutions calculated as homework.

Second, some practical problems dealing with the operation of mechanisms are solved by using the simulation software in order to understand the dynamics of these mechanisms. The calculation of the driving torque for the operation of a landing gear by solving the inverse dynamics problem and the balancing of the powertrain of a four-stroke internal combustion engine are shown as examples.

In both cases, there are some side benefits in addition to the achievement of the objectives of the lab sessions. This proposal shows to be very attractive for the students, who learn the basics of a commercial engineering tool and have a working experience closer to real life. So, they usually appreciate this methodology, leading to a better teaching experience [5].

There also some drawbacks regarding the complexity of the program, due to the great number of options available for the user and the need of an accurate modeling of the mechanisms. Some difficulties can appear when creating these exercises in order to make them robust enough for the students to build correctly the models and get the results expected.

In the following sections, some examples of the exercises developed in this virtual lab are presented and the difficulties found are pointed out.

2 Fundamentals of Spatial Kinematics

At the beginning of the course in the Aeronautical Engineering degree, some concepts related to kinematics are reviewed. For instance, the characteristics of the trajectory of a particle are studied by analyzing the curve described by a sphere with an imposed motion. Also, a practical application of the spatial kinematics equations for the experimental determination of the angular velocity of a rigid body is presented.

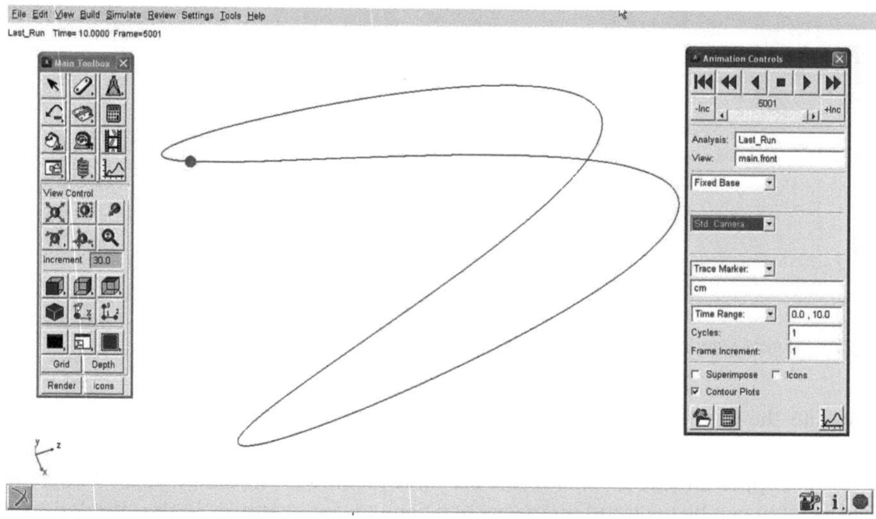

Fig. 1 Spatial trajectory

2.1 Characteristics of a Trajectory

A small sphere is created and a *General Motion* tool is used in order to prescribe its trajectory with a vectorial function of time. Then, the calculation of the Frenet frame vectors is programmed through *Requests*. The students have to compare the results given by the simulation with the analytical solution of the curve solved by means of math software. The following formulas are used:

$$\vec{u}_t = \frac{\vec{r}'}{|\vec{r}'|}\,; \quad \vec{u}_b = \frac{\vec{r}' \times \vec{r}''}{|\vec{r}' \times \vec{r}''|}\,; \quad \vec{u}_n = \frac{\left(\vec{r}' \times \vec{r}''\right) \times \vec{r}'}{|(\vec{r}' \times \vec{r}'') \times \vec{r}'|} \text{ or directly, } \vec{u}_n = \vec{u}_b \times \vec{u}_t \tag{1}$$

where $\vec{r}, \vec{r}', \vec{r}''$ refer to the position vector of the particle and the first and the second derivatives with respect to time, respectively. In Fig. 1 the trajectory can be seen and in Fig. 2, the results for the tangent vector are shown as an example of the calculations.

2.2 Experimental Determination of the Angular Velocity of a Rigid Body

There are some applications, e.g. biomechanics, where the experimental determination of the angular velocity of a rigid body can be of interest. The angular velocity of a body can be determined by measuring the velocity of three non-aligned points on its surface. Then, an over-determined linear system must be solved in order to

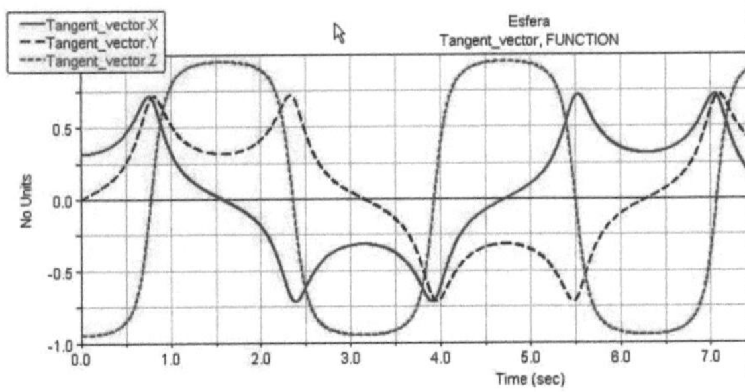

Esfera Tangent_vector, FUNCTION			
Time	Tangent_vector.X	Tangent_vector.Y	Tangent_vector.Z
0.0	0.3162	0.0	-0.9487
0.002	0.3162	4.9806E-004	-0.9487
0.004	0.3162	0.0011	-0.9487
0.006	0.3162	0.0016	-0.9487
0.008	0.3162	0.0022	-0.9487
0.01	0.3163	0.0028	-0.9487
0.012	0.3163	0.0035	-0.9487
0.014	0.3163	0.0041	-0.9487
0.016	0.3163	0.0047	-0.9486
0.018	0.3163	0.0054	-0.9486
0.02	0.3164	0.006	-0.9486
0.022	0.3164	0.0066	-0.9486

Fig. 2 Tangent vector components for the given trajectory versus time

calculate the angular velocity: there are six kinematic equations to find the three components of the angular velocity vector.

In real applications, some *markers* are distributed on the surface of the body and their motion can be captured by means of a camera, for example. In the simulation, a plate and two links are created and joined through revolute joints with different axis of rotation, as can be seen in Fig. 3. Three different angular motions are applied to the joints between the three parts, and the velocity of three markers located at the corners of the plate is measured. In Fig. 4, the trace of the three markers is shown, together with the time evolution of the velocity of one of them.

3 Dynamics of Mechanisms

When the students have a reasonable knowledge on kinematics and dynamics of a rigid body, mechanisms are introduced. The dynamics of mechanisms are presented through a series of exercises that the students have to complete from the modeling to the analysis of the simulation results.

The following subsections show two examples of this methodology.

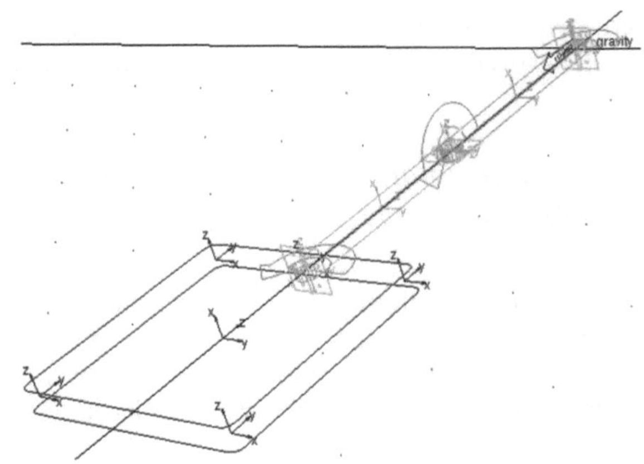

Fig. 3 Model for the determination of the angular velocity of a body

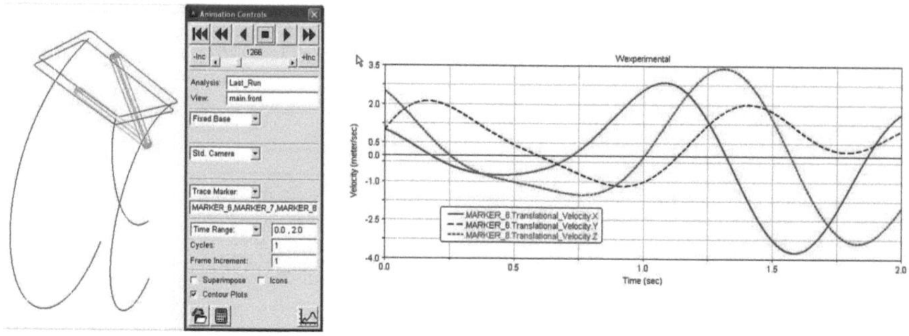

Fig. 4 Trace and velocity of a marker located at one corner of the plate

3.1 Driving Torque for the Operation of a Landing Gear

The exercise proposed in this lab session consists on calculating the driving torque necessary for the mechanism of a landing gear to complete the folding operation. This operation should be completed within a determined time, t_f, and the angular displacement of the driving link should ensure the smooth operation of the landing gear during the start and end of the folding motion. For this purpose, the angular displacement follows a control law that consists of an initial section of parabolic evolution with zero initial speed for the smooth start of the motion, then a constant speed section, and finally a parabolic section for the smooth braking of the mechanism.

In Fig. 5, the model of the landing gear and the angular displacement of the driving link are shown. The mechanism is driven by a revolute motor that joins link 2 to the ground. Students have to build the model, to program the control law in the *imposed motion* field of this joint, and to measure the driving torque necessary to achieve this motion, as shown in Fig. 6.

Fig. 5 Angular displacement of link 2 during the folding operation of the landing gear

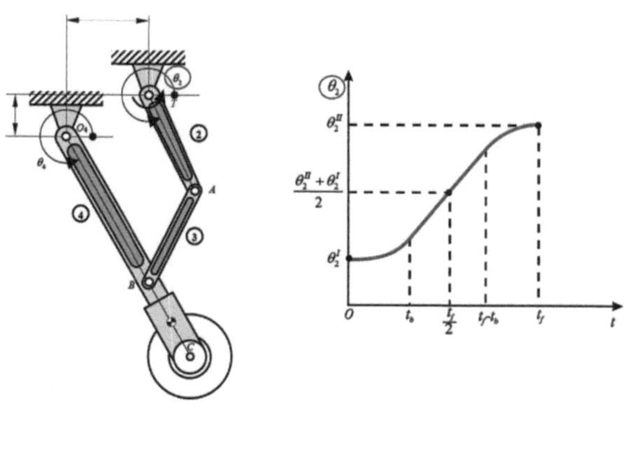

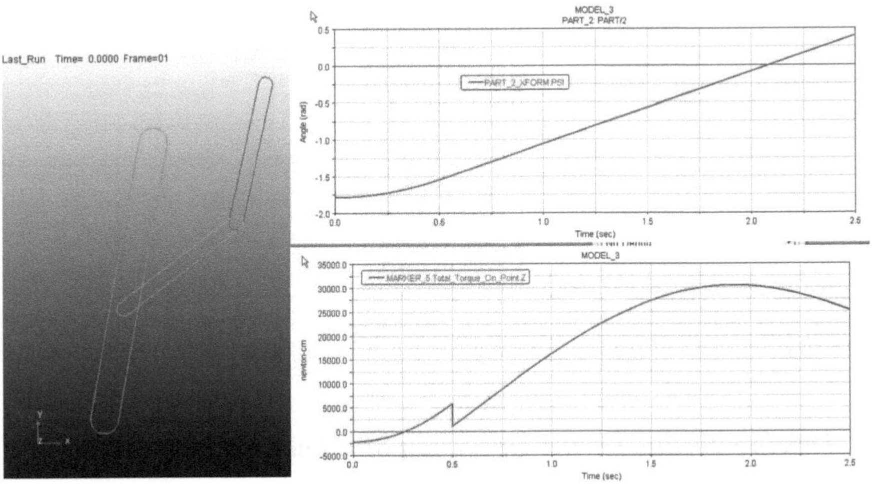

Fig. 6 ADAMS© simulation of the landing gear operation

3.2 *Balancing of the Powertrain of a Four-Stroke Engine*

This exercise has two main objectives: first is to learn how to import solids from a CAD program and how to locate them in order to make the assembly as easy as possible. A single-cylinder engine is built by using parts already available. Then, from this single-cylinder engine, students learn how to build 2, 4 or 6 cylinder engines by grouping and copying parts.

Second objective is the study of the balancing of these powertrains. Force and torque reactions on the supports can be precisely measured and different solutions to this problem analyzed.

In this case, some difficulties appear when dealing with solids imported from CAD programs. It is very important to have a precise knowledge of the geometrical characteristics of these solids in order to avoid problems when creating the joints

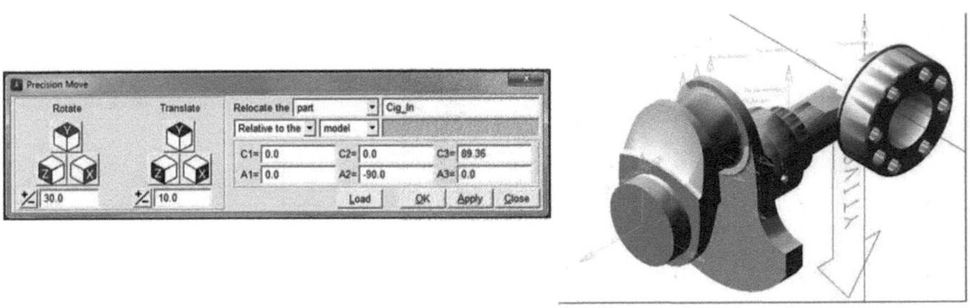

Fig. 7 Precision move tool for CAD file import in ADAMS

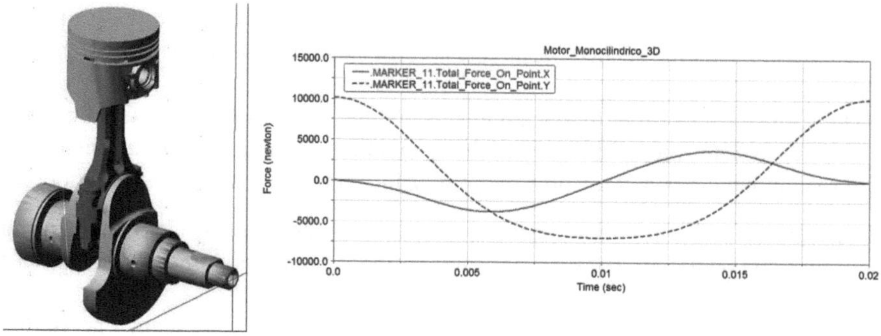

Fig. 8 Reaction forces in single-cylinder engine

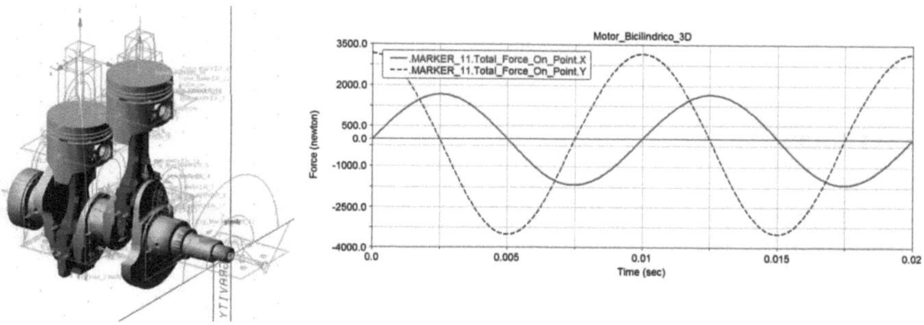

Fig. 9 Reaction forces in two-cylinder engine

between the parts of the model. The use of the Precision Move tool (see Fig. 7) is indispensable. Also, the use of the wired view is not recommended as these solids usually have lots of facets and edges.

The reaction forces in the support of a single-cylinder engine and of a two-cylinder engine are shown in Figs. 8 and 9, respectively. Y direction is the vertical direction where alternating motion of the piston occurs, and X direction refers to horizontal direction perpendicular to rotation axis. Crankshaft is not balanced initially, so X component of the force appear in both cases.

4 Conclusions

A set of exercises in ADAMS© has been developed in order to improve the learning outcomes of the students in courses related to Mechanism and Machine Science fundamentals. The advantages of using these exercises during the lab works can be summarized as follows:

- The visualization of the spatial kinematics of a particle and a rigid body, and the calculation of some related variables, help the students to better understand the theory.
- The simulation of some real mechanisms and the analysis of the results obtained give the students the opportunity to think about the dynamic problems that appear in real applications and to solve them by using a commercial engineering tool.

Moreover, the validation of the numerical results of the simulations by comparing them with the analytical solution obtained by the students as homework, improves their skills for solving exam problems.

On the other hand, some difficulties have been found. The main one is that ADAMS© working environment is not very friendly for the students, so it is difficult for them to start working with the program. It is worth to spend some initial hours learning how to manage in this environment.

References

1. Chaturvedi, S.K., Dharwadkar, K.A.: Simulation and visualization enhanced engineering education development and implementation of virtual experiments in a laboratory course - American Society for Engineering Education, AC 2011–742 (2011)
2. Fu, T.: Applications of computer simulation in mechanism teaching. Inc. Comput. Appl. Eng. Educ. **11**, 156–165 (2003)
3. Mateo Sanguino T.J., Andújar Márquez J.M.: Simulation tool for teaching and learning 3D kinematics workspaces of serial robotic arms with up to 5-dof. Inc. Comput. Appl. Eng. Educ. 20, 750–761 (2012)
4. Nickerson, J.V., Corter, J.E., Esche, S.K., Chassapis, C.: A model for evaluating the effectiveness of remote engineering laboratories and simulation in education. Comput. Educ. **49**, 708–725 (2007)
5. Suñer, J.L., Mata, V.: Utilización de programas de simulación en la docencia de la ingeniería mecánica.In: Actas del X Congreso de Innovación Educativa en las Enseñanzas Técnicas, Valencia, Spain (2002)

Using the ThinkMOTION Project Resources for the Teaching of Mechanism and Machine Theory

V. Petuya, C. Pinto, J. Corral, E. Amezua and A. Hernández

Abstract The thinkMOTION project, as a part of Europeana (www.europeana.eu) leads to the world's largest free access digital library in the field of motion systems, which is very useful for all technical subjects and in particular for Mechanism and Machine Theory. The project helps to make a significant improvement in the quantity and quality of these digital contents establishing a new kind of digital library in Europeana with focus on technical knowledge. The collected material is processed and presented immediately in a multilingual interactive portal via Europeana. Hence, the techno-cultural heritage and the current developments in motion science are preserved and made accessible for a wide range of European user groups as scientists, lecturers or students. The provided interactive material leads to a deeper understanding and motivates the students to learn more about the motion systems and machine design approaches. All kinds of material are considered (e.g. books, journals, drawings, images, physical models) to establish a digital library that connects historical and recent content from different countries. Six content providers spread over Europe are part of the project consortium. The open access to high quality material of historical and recent content from different countries opens new possibilities in multilingual searching, browsing and using of information sources.

1 Introduction

The thinkMOTION project establishes a new sector for Europeana with focus on technical knowledge which is stored in very heterogeneous forms like text documents, photos, videos, animations, technical drawings, calculation sheets, physical models and so on. Previous digitisation projects often neglected the technical, techno

V. Petuya (✉) · C. Pinto · J. Corral · E. Amezua · A. Hernández
Department of Mechanical Engineering, University of the Basque Country UPV/EHU,
Alameda de Urquijo s/n, 48013 Bilbao, Spain
e-mail: victor.petuya@ehu.es

J. C. García-Prada and C. Castejón (eds.), *New Trends in Educational Activity in the Field of Mechanism and Machine Theory*, Mechanisms and Machine Science 19,
DOI: 10.1007/978-3-319-01836-2_25, © Springer International Publishing Switzerland 2014

historical and techno cultural knowledge because nontechnicians decide which kinds of content are important to preserve for future generations. Technical knowledge tends to be buried in oblivion although this knowledge is an inseparable part of our European self-concept.

Our European cultural heritage is defined by technical developments and comforts. thinkMOTION has been the first project which allows the connection and publication of forgotten treasures of the heterogeneous content to honour the creative genius of the countless European inventors, engineers and natural scientists, which have enabled and expedited the technical progress in medicine, electrical, civil, automotive engineering etc. Our high standard of living is based on these achievements.

Six European partners have located and selected the content providers, processed the different materials and supported the multilingualism of the metadata in at least seven languages (German, English, Italian, French, Spanish, Romanian), which also complying with the Europeana requirements. The partners of the Project are: Ilmenau University of Technology (Germany), RWTH Aachen, University (Germany), University of Cassino and Southern Lazio (Italy), University of the Basque Country (Spain), French Institute of Advanced Mechanics (France) and Politehnica University of Timisoara (Romania).

Knowledge about mechanisms and machines that comes from European countries is currently difficult to access for students and researchers. Historical books in mechanical engineering are of low availability. Sources such as private or educational collections of physical models are usually not open to the public. But even the access to public content does not comply with today's requirements concerning a rapid information retrieval. The heterogeneous sources that represent the knowledge about motion systems are widespread over a lot of institutions (museums, libraries and mainly universities) and persons in nearly all European countries. Many of these sources are nonregistered material and hence not findable.

Therefore the first challenge of the thinkMOTION project has been the collection and processing of these contents in order to make them accessible for a wide range of European user groups as scientists, lecturers or students. The provided interactive material to a deeper understanding and should motivate the students to learn more about the motion systems and machine design approaches.

In this paper some of these contents will be shown in order to give to the potential users an idea of the scope of the project and the quality, quantity and variety of the contents and their potentiality to be used by Universities, students and even schools during the lessons and the personal research.

2 Processing Information to Europeana

As said, the thinkMOTION project is specialised on gathering specific content in the field of motion systems. In order to upload the different items into the database, a web application called proDB has been created [1]. Once all the information has been processed, it will be available on the internet for all the users. Currently, it is available on the Digital Mechanism and Gear Library [2], but it is being migrated to Europeana [3].

In Fig. 1 the different types of contents available in the digital portal are depicted. In Literature there are currently more than 21,000 text documents covering not only historical documents but also more recent PhD. Theses and research documents. In the Mechanism Descriptions section there are more than 1,500 mechanisms classified into different categories. In the Biographies section there are more than 3,000 bios of engineers related to the historical developments of Mechanism and Machine Theory. There are also 1,200 interactive animations and more than 30,000 commented images of mechanisms, machines and other mechanical devices. All this information is linked so for example if you are reviewing some text document, you will find a hyperlink to the bio of the author and other links to several selected images of the document.

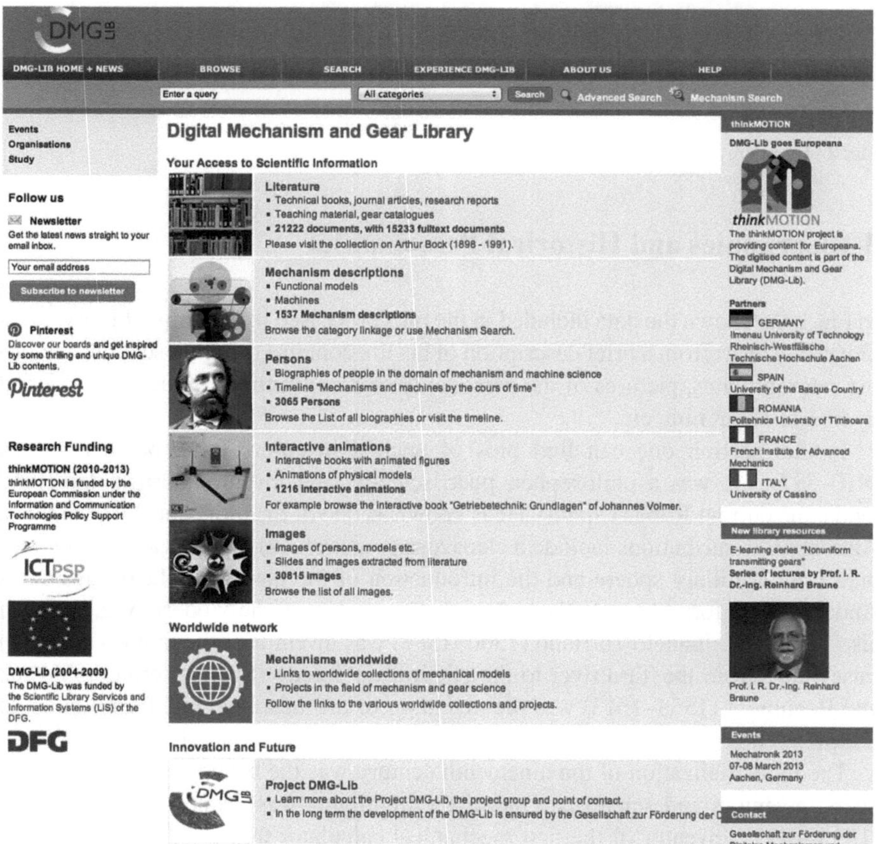

Fig. 1 DMG-Lib portal (www.dmg-lib.org)

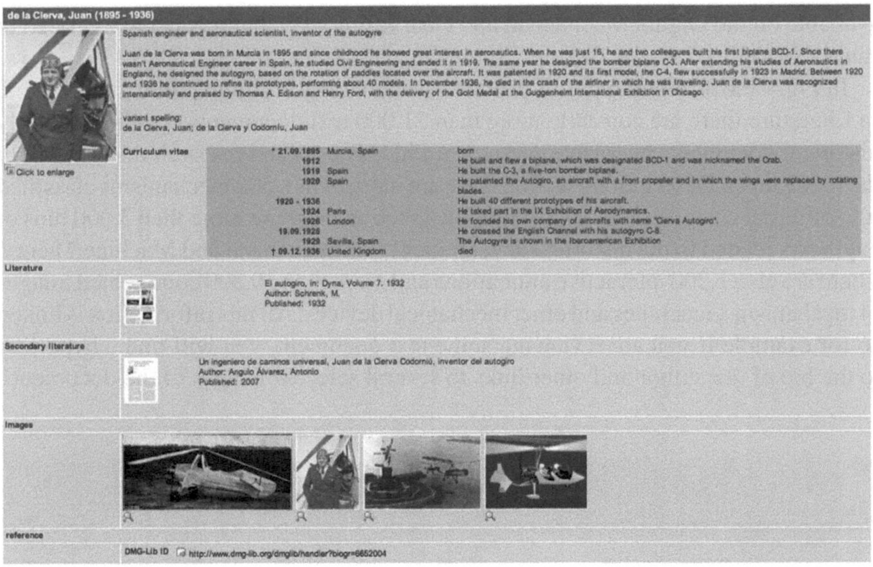

Fig. 2 Biography of Juan de la Cierva

3 Biographies and Historical Personalities

In Fig. 2 are shown the data included in the bios processed in the digital library. Each biography apart from a brief description of his life contains different sections including relevant dates, pictures of the personality or his inventions, literature produced by him or about him, etc.

In this section one can find bios of ancient inventors as Abbás Ibn Firnás (810–887). He was a philosopher, poet, scientist and inventor born in Izn-Rand Onda, the current Ronda (Spain). He is known as one of the pioneers of aeronautics. Among his contributions include a clepsydra (water clock) that he gave to Muhammad I, an armillary sphere and the introduction of the glass manufacturing in Al-Andalus [4]. Information of other relevant personalities of the Modern Age, has been also processed. Juanelo Turriano (1500–1585) was inventor of the famous device to raise water from the Tajo river to the Alcázar de Toledo (Spain). Jerónimo Ayanz and Beaumont (1553–1613) was the precursor of the steam engine. Bartolomeu de Gusmão (1685–1724) who starred the first public demonstration of a hot air balloon.

The industrialization of the nineteenth century was the birthplace of many engineers, inventors and scientists. One of the highlights was Ramon Verea Silvestre (1833–1899), inventor of the first mechanical calculator that did directly multiplications. The U.S. patent office accepted his patent on September 10, 1878 [5], and that same year he won a medal at the World Inventions Exposition in Cuba. In Fig. 3 some images of this patent are shown.

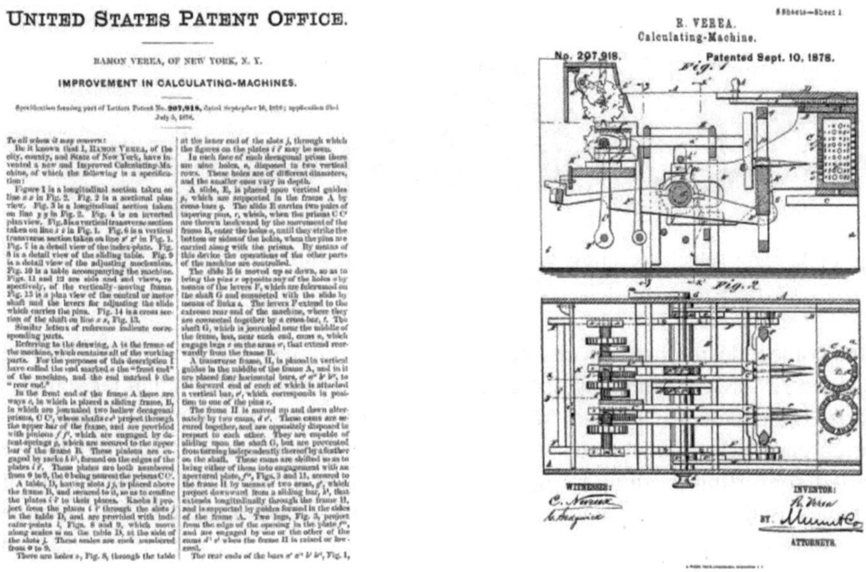
UNITED STATES PATENT OFFICE.

RAMON VEREA, OF NEW YORK, N. Y.

IMPROVEMENT IN CALCULATING-MACHINES.

Fig. 3 Ramon Verea's mechanical calculator

Other Spanish inventors of these years and whose biography is accessible in the digital portal include, Alzola and Minondo Paul (1841–1912), Narciso Monturiol Estariol (1819–1885) and Isaac Peral y Caballero (1851–1895). In the late nineteenth and early twentieth centuries it must be highlighted the figure of the Spaniard engineer Leonardo Torres Quevedo (1852–1936). One of his greatest inventions was the Telekino, a robot that executed orders sent by wireless telegraphy, being a pioneer in the field of remote control. Other of his remarkable inventions are an electromechanical machine to solve the algebraic equations, a programmable chess player, a new type of dirigible, a cable car at Mount Ulia in San Sebastian (Spain) and the Niagara Aerocar at the Niagara Falls (USA) [6].

In the early twentieth century, figures like Juan de la Cierva Codorniú (1895–1936) and Federico Cantero Villamil (1874–1946), promoted the rotary wing flight with their inventions. The first invented the autogyro, a hybrid of airplane and helicopter (not invented yet) based on the rotation of blades located on the aircraft. It was patented in 1920 and its first model, the C-4, flew successfully in 1923 in Madrid. Meanwhile, Federico Cantero Villamil worked on the construction of the helicopter. After some failed attempts, in 1935 he began construction of his helicopter, the Libélula Viblandi, but it was delayed due to the Spanish Civil War. In 1941, the helicopter was ready for flight-testing, but eventually fell into oblivion after the successful flight test conducted by Igor Sikorsky in 1939 [7].

The twentieth century also left us two important personalities in the Spanish automotive industry: Wifredo Ricart (1897–1974) and Eduardo Barreiros (1919–1992). The first was an engineer who worked on the companies Hispano-Suiza and

Fig. 4 Barreiros vehicles (courtesy of Fundación Barreiros)

Alfa Romeo, designing from engines for aviation to engines for racing cars. Later he founded ENASA, a company dedicated to the manufacture of commercial vehicles whose trademark was Pegaso. Meanwhile, Eduardo Barreiros was a mechanic and entrepreneur who contributed significantly to the improvement of diesel engines. He started converting gasoline to diesel engines in order to meet the special needs of vehicles used in his company and later he founded Barreiros Diesel SA. He became to associate with Chrysler Corporation but finally he lost the control of his own company. Figure 4 shows images of some of his vehicles.

4 Digitisation of Historical Documents

The project also has been responsible for collecting and digitizing valuable historical documents. As mentioned in the introduction, one of its objectives is to make accessible the techno-cultural heritage and one of the major sources of such heritage are precisely the old volumes of books and journals. Scanned documents pass a cleaning process before being imported into the system. After scanning, a cleaning process is carried out. Using Scantailor software, scanned files are aligned and focused being eliminated possible stains or marks due to the age. Finally, using the text recognition program ABBY FineReader, the pages become recognized text PDF files.

An interesting book is El Goniobarímetro by Dario Bacas (1845–1913) [8]. Dario Bacas was an engineer that spent much of his life in Bilbao, being one of the founders and the first director of the Faculty of Engineering in Bilbao. This book describes a device used to measure the weight based on a particular property of the cycloid curves. It has been digitized and is now available to be read in the digital portal. Figure 5 shows the cover of El Goniobarímetro before the cleaning process and the final result as pdf file.

Other remarkable book is Machinae Novae by Fausto Verantio in 1615 in which one can find different mechanical devices as a gearwheel mill shown in Fig. 6 used in Italy and Greece and generally moved by donkeys or horses.

Besides books, old volumes of technical journals have been also processed for the digital library. After an agreement with the Spanish Association of Mechanical Engineering, the Annals of the National Congress of Mechanical Engineering have

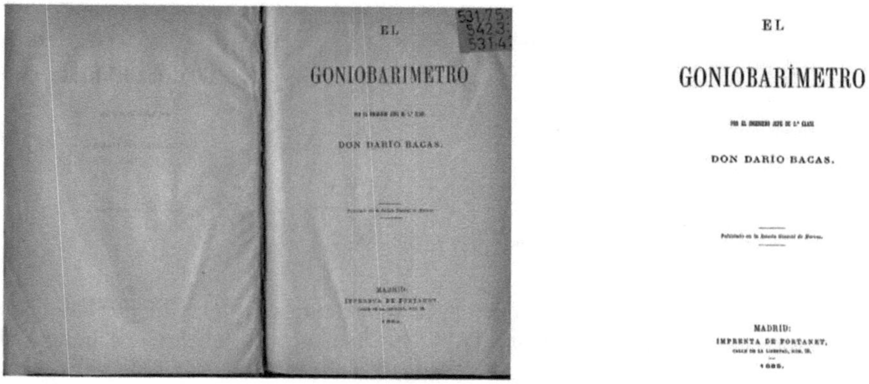

Fig. 5 El Goniobarímetro (Darío Bacas)

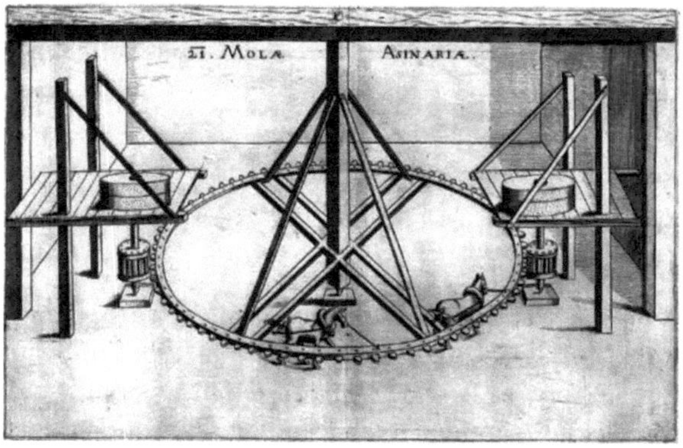

Fig. 6 Machinae Novae: gearwheel mill

been incorporated. All volumes since the first Congress in 1982 have been digitized and are now available online. Another source of interesting documents is DYNA journal. The Federation of Industrial Engineers of Spain publishes this journal. The first issue was published in January 1926 with 500 copies. Since then, the magazine has been published continuously until today. In its pages, one can find articles on subjects as the fall of the Tacoma bridge, the inventions of Juan de la Cierva or the celebration of the centenary of the Mechanical Engineering degree. There are also available many articles on combustion engines and railways, by relevant authors as Alejandro Goicoechea Oriol, inventor of the Talgo train.

5 Mechanism Descriptions

As said, the thinkMOTION is not just about collecting relevant historical documents, but also to locate machines and mechanisms and introduce their descriptions in the database. The mechanisms can be both current mechanical systems and older

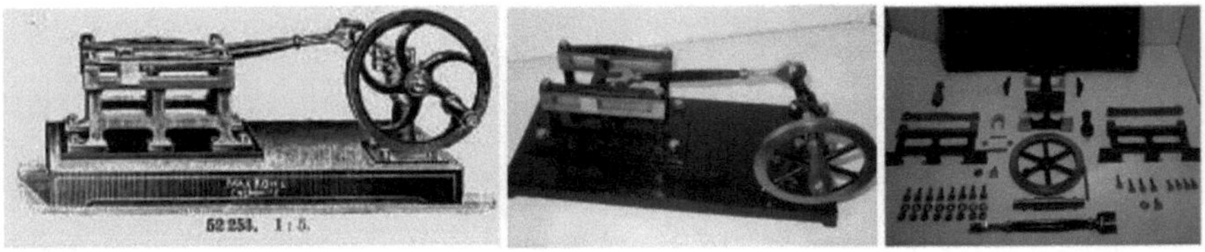

Fig. 7 Max Kohl: slider-crank linkage

machines and prototypes still in operation. Interesting examples of the latter are the prototypes manufactured by Max Kohl [9] located in the School of Industrial Engineering (Béjar) of the University of Salamanca (Fig. 7).

For each prototype, the library offers the user detailed and useful information of the mechanism, such as the functionality of the mechanism, the structure of the kinematic chain, number of elements, outputs, inputs, degrees of freedom, etc. Photographs, diagrams, videos or interactive animations of the mechanisms can be also found in the library.

6 Conclusions

This paper presents a description of the work being done in the European project thinkMOTION. It presents the various types of documentation that is stored in the digital library, showing some of the most significant examples. The research being done has also brought to light findings of historical personalities who made significant advances in mechanical engineering. Through this article we want to disseminate results of the project and promote the use of the digital portal for teaching a research purposes by professors, students and researchers.

Acknowledgments The authors wish to acknowledge the financial support received from the Spanish Government through the Ministerio de Economía y Competitividad (Project DPI2011-22955), the European Union (Project FP7-CIP-ICT-PSP-2009-3) and the Regional Government of the Basque Country through the Dpto. Educ, Univ. e Investig. (Project IT445-10) and UPV/EHU under program UFI 11/29.

References

1. Brix, T., Döring, U., Reeßing, M.: Creating present-day solutions from historical knowledge. In: 13th World Congress in Mechanism and Machine Science, Guanajuato, México, 19–25 June, (2011)
2. Digital Mechanism and Gear Library: http://www.dmg-lib.org
3. Europeana: http://www.europeana.eu
4. Rius, M.: El sabio Total: Ibn Firnás. Revista Jábega **97**, 9–13 (2008)
5. Verea, R.: Improvement in calculating-machines. US Patent No. 207.918 (1878)

6. Torres-Quevedo Torres-Quevedo, L.: Leonardo Torres Quevedo. Ingeniería y Territorio **79**, 54–57 (2007)
7. Díaz de Aguilar Cantero, I., Suárez Caballero, F.: Federico Cantero Villamil. Entre la desmemoria y el revisionismo, Ingeniería y Territorio **79**, 64–73 (2007)
8. Bacas, D.: El Goniobarímetro. Imprenta de Fortanet, Madrid (1885)
9. Kohl, M., Chemnitz, A. G.: Price List No. 50, Vols II and III. Physical Apparatus. Chemnitz (1911)

Experience of Modernization of the Curriculum TMM in St. Petersburg State Polytechnical University

Alexander Evgrafov, Andrey Khisamov and Olga Egorova

Abstract At the end of twentieth century, the subject of the theory of mechanisms and machines greatly expanded. Multi-engine machines with open and closed kinematic chains become widespread. Requirements to precision of mechanisms increased. PCs became widely used in the learning process. All this resulted in the need to modernize the course on the theory of mechanisms and machines. This paper describes the modernization of the discipline "Theory of mechanisms and machines" undertaken at the St. Petersburg State Polytechnic University [1]. Modernization encompassed structural, geometric, kinematic and force analysis of mechanisms and affected the problems concerning the dynamics of machines [2].

Keywords Theory of Mechanisms and Machines · Modernization · SPSPU

1 Structural Analysis of Mechanisms

New type concepts were based on the following definition of the structural group.

A structural group (or type group) is a kinematic chain with a given location of inputs, where the number of inputs is equal to the number of degrees of movability [3]. The chain links are assumed to be rigid, and in determining the number of degrees of

A. Evgrafov (✉) · A. Khisamov
St. Petersburg State Polytechnical University (SPbSPU), St. Petersburg, Russia
e-mail: a.evgrafov@spbstu.ru

A. Khisamov
e-mail: andrey@k-digital.ru

O. Egorova
Bauman Moscow State Technical University (BMSTU), Moscow, Russia
e-mail: cuba2006@inbox.ru

J. C. García-Prada and C. Castejón (eds.), *New Trends in Educational Activity in the Field of Mechanism and Machine Theory*, Mechanisms and Machine Science 19,
DOI: 10.1007/978-3-319-01836-2_26,

movability the internal and external inputs of the kinematic chain and links imposed by both internal and external kinematic pairs are taken into account.

Such definition allows us to include in the subject matter of the TMM almost all existing mechanisms [4]. Note that it is still possible to use the known methods of study (though some adaptation of these methods is required sometimes) [5].

The concept of a simplest structural group is introduced; the simplest structural group cannot be divided into several simpler groups. If the number of degrees of movability and the number of inputs are zero, the structural group is the ordinary Assur group.

When consecutively connecting structural groups to a frame, at each stage, a mechanism is created in which the number of degrees of movability w is equal to the number of inputs n; such a mechanism is called *normal*. This concept of structural group allows us to describe the type of any normal mechanism; thereby, a design model of a mechanism may be formed suitable for its geometric, kinematic and dynamic analysis, because all stages of the analysis may be carried out sequentially and separately for each structural group.

Thus, it is becomes possible to describe a mechanism by a *type graph (structural graph)* in which vertices are structural groups and ribs are kinematic pairs connecting them. For example, Fig. 1 shows the aerial platform: a—the diagram of the mechanism, b—the type graph.

The first digit in the vertices is the number of links of the group, and the second digit is the number of degrees of movability of the group.

Special attention is paid to the *mechanisms with redundant constraints*. The structural analysis allows categorizing all constraints created by kinematic pairs into two groups. A constraint the deletion of which will not change the number of degrees of movability is called *non-releasing*, while a constraint the deletion of which will increase the number of degrees of movability per unit is called *releasing*. Such categories are useful, for example, in the dynamic force analysis. It turns out that in a statically determinable mechanism (with redundant constraints), the releasing constraint forces can be determined, while the non-releasing constraint forces cannot be determined.

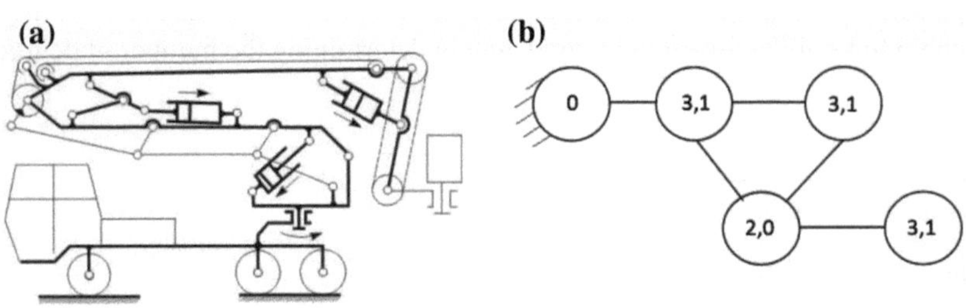

Fig. 1 Aerial platform: **a**—the diagram of the mechanism, **b**—the type graph

2 Geometric Analysis of Mechanisms

The geometric analysis studies the position functions of mechanisms which, in case of linkage, may be represented in the form (1), where $q_1, \ldots, qn$—generalized input coordinates, $\alpha_1, \ldots, \alpha_m$—parameters of mechanism.

$$x_k = \Pi_k(q_1, \ldots, q_n, \alpha_1, \ldots, \alpha_m), k = 1, \ldots, N \tag{1}$$

The geometric analysis is associated with a number of problems caused by the non-linearity of Eq. (1) and, in most cases, by the inability to represent them in the explicit form. Functions (1) are often multi-valued. In some cases, this is due to the existence of several different assemblies, in other cases, due to the existence of several possible positions for the set generalized coordinates. In the latter case, the transition from one possible position to the other is accompanied by a passage through a special position of mechanism, which is generally undesirable. This entails the need to analyze specific positions and to determine the conditions for the uniqueness of the solution of a geometrical problem.

In order to solve the problem of the geometric analysis, the following procedure is proposed:

- The geometric analysis is performed sequentially for each of the structural groups, starting from the frame. Each subsequent group may be adjoined either to the frame or to any preceding group, according to the type diagram of mechanism. The inverse problem is solved similarly, but for the inverted type mechanism.
- For each group, the input and output coordinates are introduced. The input coordinates are generalized coordinates q_s of the mechanism that fall into this group, as well as the coordinates that define the position of the kinematic pairs of the preceding groups to which the group in question is joined. The output coordinates are the coordinates that define the position of the kinematic pairs joined by subsequent groups, and the output coordinates of the mechanism.
- The set input coordinates of a group do not uniquely define its output coordinates. To resolve this ambiguity, additional "*group*" coordinates are introduced and equations compiled that link the group coordinates with input and output coordinates. To this end, in the group under consideration, some kinematic pairs are opened to represent the group type as a tree. Next, the coordinates are introduced that define the position of each next link relative to the previous one. They may include the input coordinates of the mechanism, and all the rest will be group coordinates. Note that the number of group input coordinates is always equal to the number of open links. Closure conditions drawn up for all open links form "*group equations*" the solution of which resolves the problem of geometric analysis.

This procedure is illustrated using a planar mechanism with three degrees of movability (Fig. 2) and input coordinates q_1, q_2, q_3 [3]. It consists of the two

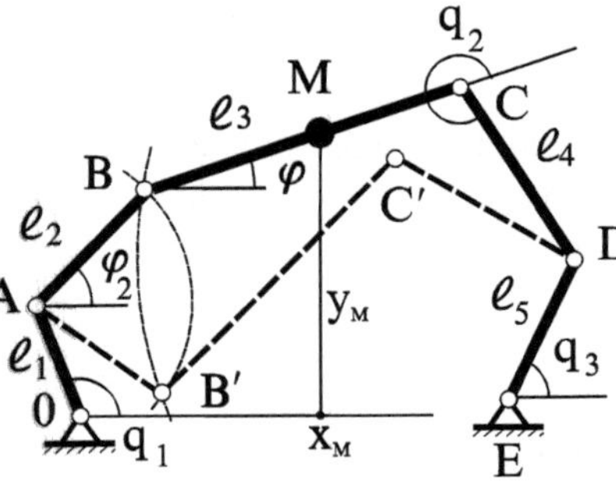

Fig. 2 A three degree-of-freedom planar mechanism

single-link one degree-of-freedom groups I and II joined to the frame and three-link one degree-of-freedom group III. The input and output coordinates are presented in the Table for each of the groups. Note that the output coordinates of the entire mechanism are x_M, y_M, φ, which define the position of the VS "platform" positioned by the three drives. Equation (2) set the relations between the input and output coordinates of groups I and II; as these coordinates feature a tree-like structure, there is no need to introduce any group coordinates.

Group	Input coordinates	Output coordinates	Group coordinates
I	x_o, y_o, q_1	x_A, y_A	–
II	x_E, y_E, q_3	x_D, y_D	–
III	x_A, y_A, x_D, y_D, q_2	x_M, y_M, φ	φ, φ_2

$$\text{I}\quad \begin{aligned} x_A &= x_0 + \ell_1 \cos q_1, \\ y_A &= y_0 + \ell_1 \sin q_1. \end{aligned} \qquad \text{II}\quad \begin{aligned} x_D &= x_E + \ell_5 \cos q_3, \\ y_D &= y_E + \ell_5 \sin q_3. \end{aligned} \tag{2}$$

In group III, by opening the joint D, we get a "tree". By entering the two group coordinates φ_2 and φ (their number is equal to the number of open links) and by projecting the path ABCD on the x and y axes, we obtain the group Eqs. (3) setting the relations between the group coordinates φ_2 and φ and the input coordinates x_A, y_A, x_D, y_D, q_2.

$$\begin{aligned} \ell_2 \cos\varphi_2 + \ell_3 \cos\varphi + \ell_4 \cos(\varphi + q_2) - x_D + x_A &= F_1(\varphi_2, \varphi) = 0, \\ \ell_2 \sin\varphi_2 + \ell_3 \sin\varphi + \ell_4 \sin(\varphi + q_2) - y_D + y_A &= F_2(\varphi_2, \varphi) = 0. \end{aligned} \tag{3}$$

Thus, the problem of geometrical analysis has been reduced to solving a system of trigonometric equations. Having solved them, let us find x_M, y_M, φ from relations (4).

$$\begin{aligned} x_M &= \ell_2 \cos\varphi_2 + BM \cos\varphi + x_A, \\ y_M &= \ell_2 \sin\varphi_2 + BM \sin\varphi + y_A,\ \varphi = \varphi. \end{aligned} \tag{4}$$

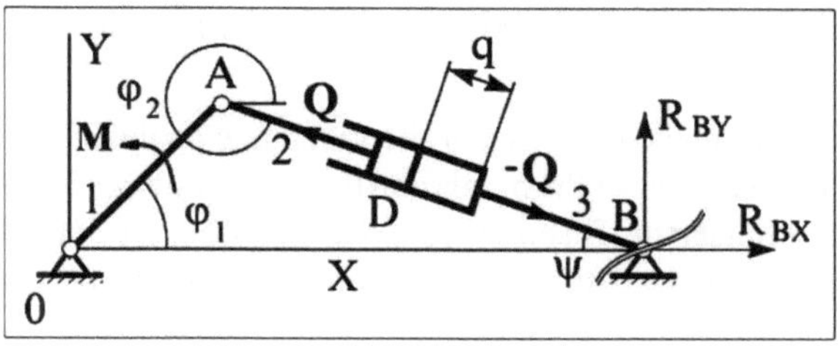

Fig. 3 Example of the mechanism

Given the input coordinates, the Eqs. (3) have two solutions; the second one is shown in Fig. 3 by a dotted line. Solving the equations using the computer, you must specify which of them is main and which is secondary. This can be done in different ways by specifying certain geometric criteria (e.g. a sign of the vector product $\overrightarrow{AB} \times \overrightarrow{BC}$). Multiple solutions of group equations mean that there is a special position in which both of these solutions are the same. From the theory of implicit functions it follows that, in the special position, Jacobian of the system (3) is zero (condition (5)). In the special position, joints A, B, D lie on the same straight line.

$$J = \begin{vmatrix} \frac{\partial F_1}{\partial \varphi_2} & \frac{\partial F_1}{\partial \varphi} \\ \frac{\partial F_2}{\partial \varphi_2} & \frac{\partial F_2}{\partial \varphi} \end{vmatrix} = \begin{vmatrix} -\ell_2 \sin \varphi_2 & -\ell_3 \sin \varphi - \ell_4 \sin(\varphi + q_2) \\ \ell_2 \cos \varphi_2 & \ell_3 \cos \varphi + \ell_4 \cos(\varphi + q_2) \end{vmatrix}$$
$$= \ell_2 \left[\ell_3 \sin(\varphi - \varphi_2) - \ell_4 \sin(\varphi - \varphi_2 - q_2)\right] = 0 \quad (5)$$

The TMM course shall include the general methods for solving group equations. We propose to familiarize the students with the following methods:

- graphic method;
- Newton method.

The inverse problem of geometrical analysis can also be solved sequentially for each structural group, but the type representation must be made for the inverse mechanism.

In the metric synthesis of a mechanism (in defining the parameters of a kinematic diagram), the external and internal conditions for the flow of forces are determined [6].

The external conditions for the flow of forces express the ratio between the external forces: the driving torque and the workload. The internal conditions for the flow of forces characterize the relationship between the internal forces (constraint forces) and the workload. For the static model, these conditions are determined by the geometrical parameters: the size of links and their relative positions.

3 Kinematic Analysis of Mechanisms

The ultimate objective of the kinematic analysis is to determine the velocities and accelerations of the points and the angular velocities and accelerations of the links in a mechanism. This problem comes down to the determination of the first and second partial derivatives of the position functions (1), because the relations (6) exist.

$$\dot{x}_k = \sum_{S=1}^{n} \frac{\partial x_k}{\partial q_S} \dot{q}_S; \quad \ddot{x}_k = \sum_{S=1}^{n} \frac{\partial x_k}{\partial q_S} \ddot{q}_S + \sum_{S=1}^{n} \sum_{r=1}^{n} \frac{\partial^2 x_k}{\partial q_S \partial q_r} \dot{q}_S \dot{q}_r. \tag{6}$$

The study of the accuracy of a mechanism comes down to the same problem, as it requires determining the partial derivatives of the position functions using the parameters in the first approximation. Position errors δx_k resulting from coordinate errors δq_S and parameter errors are determined by the formula (7).

$$\delta x_k = \sum_{S=1}^{n} \frac{\partial x_k}{\partial q_S} \delta q_S + \sum_{l=1}^{n} \frac{\partial x_k}{\partial \alpha_l} \delta \alpha_l. \tag{7}$$

The kinetic analysis can be performed sequentially for each of the structural groups, starting from the frame. The equations for determining the first partial derivatives can be obtained by direct differentiation of group equations. To determine the derivatives for the group coordinates for a parameter α_ℓ, we differentiate the group equations for this parameter.

For planar linkages, the graphic-analytical method can be applied.

4 Force Analysis of Mechanisms

Identification of constraint forces in the kinematic pairs and generalized driving force is the traditional problem in the force analysis of mechanisms. This problem is solved through kinetostatics equations presented in one form or another. When redundant links are present, the problem becomes statically indeterminable. Some of the force analysis procedures "conceptually" associated with the above methods of geometric and kinematic analysis are considered below.

1. Let the mechanism without redundant links and with ideal kinematic pairs (frictionless) contain N movable links and p_S s-movable pairs. Then, the number of its degrees of movability w is determined by the formula (8).

$$w = 6N - \sum_{S=1}^{5} (6 - S) p_S \tag{8}$$

In this formula, the number of unknowns is equal to $6N$ which is the total number of kinetostatics equations. The same correlation between the number of unknowns and the number of equations is in place for each of the structural groups, assuming that the constraint forces are known in the "point of joint" of the following group to a group under consideration. Therefore, we can make a force analysis of the mechanism sequentially, starting from the last joined group.

2. We offer the following force analysis procedure for the structural group, which we consider on the example of the mechanism shown in Fig. 3.

   - The group AB is opened and becomes a tree (Fig. 3). The constraint forces (R_{Bx}, R_{Ay}) are introduced for the open pairs.
   - The group coordinates (φ_1, φ_2) are introduced; together with the input coordinates (q), they form a system of coordinates which uniquely determine the positions of all links in the tree.
   - The equilibrium equations are composed in the form of the moment equations about the axes corresponding to the angular coordinates of the tree (joints 0, A) and the equations of forces about the axes corresponding to the linear coordinates. The result is a system of equations, from which the constraint forces of the open links and the generalized driving forces (Eqs. (9)) are determined.

$$
\begin{aligned}
&-R_{BX}(y_B - y_0) + R_{BY}(x_B - x_0) + \sum_i M_0(\overline{P}_i) + \sum_i M_0(\overline{\Phi}_C) + M = 0;\\
&-R_{BX}(y_B - y_A) + R_{BY}(x_B - x_A) + \sum_i M_A(\overline{P}_i) + \sum_i M_A(\overline{\Phi}_C) = 0;\\
&R_{BX}\cos\psi - R_{BY}\sin\psi + Q + \sum_i (P_i)_{DB} + \sum_i (\Phi_i)_{DB} = 0.
\end{aligned}
\tag{9}
$$

where: $M_0(\overline{P}_i)$ are the moments of active forces about the axes O and A, M is the workload (moment).

To determine the drive moment we can also use the general equation of dynamics. Same results obtained by the two methods indicate the correctness of the calculations.

5 Difference in Curricula

The master's graduate degree program is expanded with the following sections:

- Dynamic force analysis for friction mechanisms: the basis of the force analysis of friction mechanisms in the kinematic pairs is, as usual, the assumption of Coulomb friction. In the case of a point contact in the pair, all the components of main vector and main moment of constraint forces can be expressed by the normal constraint force. However, the lower pairs require computational models to be specified taking

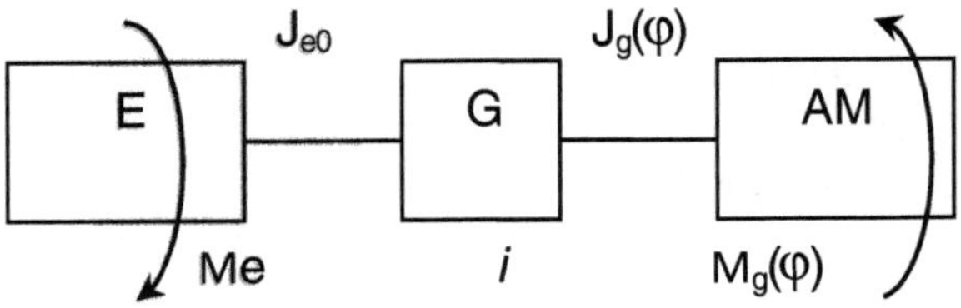

Fig. 4 Dynamic. Steady mode. E—engine, G—gear (reducer), AM—actuating mechanism

into account the additional assumption about the distribution of the normal forces on the contact surfaces. Kinetostatics equation obtained for friction mechanisms are non-linear. They may have several solutions or no solution at all.

- Analysis of elastic linkages: *elastic linkage* or *elastic mechanism* is a physical model of an actual mechanism obtained in the assumption that some links and kinematic pairs comprising the mechanism are deformable. Elastic linkages may be obtained from rigid linkages by replacing some rigid links with elastic ones. This conversion is possible by a variety of ways, depending on which links are assumed to be elastic. Static problems relating to elastic mechanisms arise, for example, in an analysis of its static errors. They include the determination of static deformations and are associated with the concept of reduced rigidity of mechanism. In the kinematic analysis of elastic linkages, a position function is considered as dependent on both the generalized coordinates $\overline{q}$ and the deformations $\overline{\theta}$.
- Dynamics: in studying the dynamics of a single-engine machine, we consider the model (Fig. 4).

The following system of differential equations is solved:

$$\begin{cases} J_g(\varphi)\ddot{\varphi} + \frac{1}{2}\dfrac{dJ_g(\varphi)}{d\varphi}\dot{\varphi}^2 = M_e + M_g(\varphi), \\ \tau\dot{M}_e + M_e = M_0 - si^2(\dot{\varphi} - \omega_0). \end{cases} \tag{10}$$

The solution is found for the two modes: steady and speeding-up. In such a way, the law of motion and the driving moment are determined.

6 Conclusion

The teaching practice at the TMM Department of the St. Petersburg State Polytechnic University shows a benevolent attitude of specialized departments to the ongoing modernization [7, 8]. Students gain experience in calculating the mechanisms of modern machines, which facilitates their success in specialized departments.

References

1. Kolovsky, M., Evgrafov, A.: O nekotorih napravleniyah modernizacii kursa TMM. In: TMM, vol. 1, 11, pp. 3–30 (2003)
2. Maunder, L.: The progress of IFToMM. Mech. Mach. Theor. **15**, 415–417 (1980)
3. Kolovsky, M., Evgrafov, A., Semenov, Yu., Slousch, A.: Advanced Theory of Mechanisms and Machines. Springer, Berlin, p. 396 (2000)
4. Evgrafov, A. Programma rascheta dinamiki odnodvigatelnoy mashini. In: TMM, vol. 1, 11, pp. 83–91 (2003)
5. IFToMM: Standardization and terminology. Mech. Mach. Theor. **38**(7–10), special issue, (2003)
6. Evgrafov, A. et al.: Teoriya mehanizmov i mashin. SPbSPU, St.-Petersburg (2009)
7. Ceccarelli, M.: The challenges for machine and mechanism design at the beginning of the third millennium as viewed from the past. In: Lectures, Invited (ed.) Proceedings of Brazilian Congress on Mechanical Engineerings COBEM2001, pp. 285–302. Dordrecht, Uberlandia, Kluwer (2004)
8. Ceccarelli, M.: Evolution of TMM (Theory of Machines and Mechanisms) to MMS (Machine and Mechanism Science): an illustration surves. Keynote Lecture, 11th IFToMM World Congress in Machine and Mechanism Science, Tianjin, vol. 1, pp. 13–24 (2004)

Part V
Mechanical Engineering Education: Experiences

The Experience of the University of Oviedo in the Motostudent Competition

A. Noriega, R. Fernández and J. L. Cortizo

Abstract This paper shows the experience of the University of Oviedo in the two editions of the international competition MotoStudent. Firstly, this university competition is described. Here, we describe the experience of one of the two teams at the University of Oviedo in the first edition indicating previous planning, project development, results in competition and identified problems. Before considering the participation in the second edition, we conducted a preliminary analysis of the technical and organizational aspects detected in the first participation of the University of Oviedo that could be improved and we proposed ways to improve them. These improvements were implemented in the participation in the second edition of the competition, which is described below, with the same approach that the participation in the first edition (planning, real development, results in competition and identified problems). Finally, some comments are made about the benefits of participating in this competition for students, teachers and universities.

Keywords Motostudent · Project-based learning · Work-based learning · Work-group skills

1 Introduction

Associations like American Society for Engineering Education (ASEE) and the National Academy of Engineering (NAE) highlight the need of changes in the STEM (science, technology, engineering, and mathematics) education. Some editors, as Engineering Education Research and Practice, expect authors to explain how their contributions represented good practice. This paper includes a comprehensive

A. Noriega (✉) · R. Fernández · J. L. Cortizo
University of Oviedo, Asturias Spain
e-mail: noriegaalvaro@uniovi.es

J. C. García-Prada and C. Castejón (eds.), *New Trends in Educational Activity in the Field of Mechanism and Machine Theory*, Mechanisms and Machine Science 19,
DOI: 10.1007/978-3-319-01836-2_27,

assessment of the various phases of hands-on learning in engineering, through the experience of a real construction of a motorcycle.

Motostudent is an international competition in which university students who have completed at least 50 % of their studies, design, manufacture and run a prototype racing motorbike. Students, guided by a tutor, have a period of three semesters to complete the project and build the prototype. It is a biennial competition that has been held in two editions: September 2010 and October 2012.

The job evaluation has two main phases. In the first one (MS1), engineering and marketing projects are valued. In the second phase (MS2), is checked whether the prototype meets all safety requirements and standards, and the motorbike is tested on track.

The organization of the teams, who have between 7 and 15 participants, simulates a manufacturer of racing motorbikes. The final design is limited by technical and economic constraints (strict regulations, the mandatory use of some commercial components, serial manufacturing of 500 units per year with unit cost less than €4500).

This competition is promoted by the Moto Engineering Foundation (MEF) that has the support of relevance Motorcycling Federations as RFME, FARAM; Motorcycles Manufacturers (ANESDOR); Championships Promoters (DORNA); Public Institutions, such as the Instituto Aragonés de Fomento, and The University of Zaragoza; Confederation of Businesses; Official College of Industrial Engineers and the Spanish Association of Mechanical Engineering.

In this paper, we will expose our experiences in the first edition, and the learning derived from it for the preparation of the second edition and how this has yielded good results. It has also allowed us to assess the impact of this activity in academic education of students [1, 2].

2 Participation in I Edition

2.1 Previous Planning

In first Edition, the planning and distribution of tasks was leave on student hands. The student team manager organized the tasks, assigning them a responsible and defining their duration. Each student was responsible for the design, the manufacture or purchase and the assembly of the components assigned to him.

The student responsible for each task should interact with his colleagues that design components connected to theirs. It led to the formation of informal working groups focused on the main components of the motorbike (engine, chassis, fairing and electronics).

In addition, each working group had to find sponsors or partners that allow them to address the fabrication or purchase of components that was responsible and helped the team expenses.

The University tutor merely supervised the work of the students and acted as technical advisor on demand.

2.2 Project Development

Students used different CAD-CAM software to design. Some programs are known because they had been used them in certain subjects (Solidworks, Fluent,...) but the students did not have enough level to solve real problems as they needed. In other cases, it was necessary to learn to use new software (ADAMS, ANSYS,...) to solve some complex calculations. In both cases, the students spent several weeks learning autonomous use of these programs.

In some cases, for instance, for the suspensions, the students spent additional time to train them in design and analysis issues that they had not developed during their bachelors. In other cases, for instance, the CDI which controls the ignition time, is also required an additional training for practical implementation of the designs.

After conclude the design and selection of the basic parts of the motorbike, the students started manufacturing some components and purchasing the commercial elements. As there was very little money and material means for this phase, students had to make many components personally (for instance, lathe cutting or welding). This meant that, in all cases, they must make an independent learning on the manufacture means available at University before starting the manufacturing process.

Despite the few material means available, students were highly motivated and were able to finish the prototype in time for the final tests.

2.3 Results in Competition

In phase MS1 (which assesses the industrial project), teams for the University of Oviedo (UniOvi) obtained the results shown in Table 1.

Considering all the scores with the weights set established by the competition organizers, in phase MS1, the TSK NRT team was ranked 11 of 24 teams and the MDU team was ranked 14 of 24 teams.

Table 1 Results for the two UniOvi teams in phase MS1 of I edition

Aspect	MDU team	TSK NRT
Design	Position 4 of 24	Position 8 of 24
Technical calculations	Position 9 of 24	Position 10 of 24
Industrialization systems	Position 18 of 24	Position 11 of 24
Cost analysis	Position 17 of 24	Position 12 of 24

In phase MS2, the prototype built by the MDU team failed the test track with the professional riders of the organization as it was not considered safe to compete in a real competition test. The prototype built by the team TSK NRT was found valid to compete but it had many problems with the engine. In the phase, TSK NRT and MDU teams were classified, respectively in positions 23 and 24 of 24 teams.

2.4 Identified Problems

Almost all students of the teams had no previous experience with motorbikes of any kind or were big fans of this type of machines. It made the designs were not, in some cases, thoughtful.

Of the original 11 students of MDU team (7 mechanical and 4 electronics), 3 withdrew from the competition although 3 new ones were captured so that the team kept the total number of members.

The scheduled task sequence was unrealistic, especially in the manufacture and, therefore, no time was available for tests before using the motorbike on a circuit.

The students had no experience in the use of machine-tools and welding equipment. This made the quality and accuracy of the parts produced was low.

The student team manager did not have enough authority to impose to other teammates, perhaps, because the hierarchy was not well established.

3 Participation in II Edition

3.1 Previous Planning

To plan the participation in the II edition, the problems identified during the participation in I edition were taken into account and some ways to avoid or minimize them were proposed.

First, the participation of the University of Oviedo was structured through only one team, the MS2-Uniovi, to combine efforts and financial resources so that they supposed the least possible burden for all involved (teachers, students, sponsors and partners).

This team would have the maximum number of members allowed in the rules of the competition (15 members) in order to good share the work and to cover more technical areas that in I edition were not enough covered (for instance, study of manufacturing systems and cost analysis in phase MS1).

The team structure was set before the selection of the students. This structure can be seen in Fig. 1.

The selection of the students to be team members was made through a competitive selection process in which interested students should introduce their candidacy by

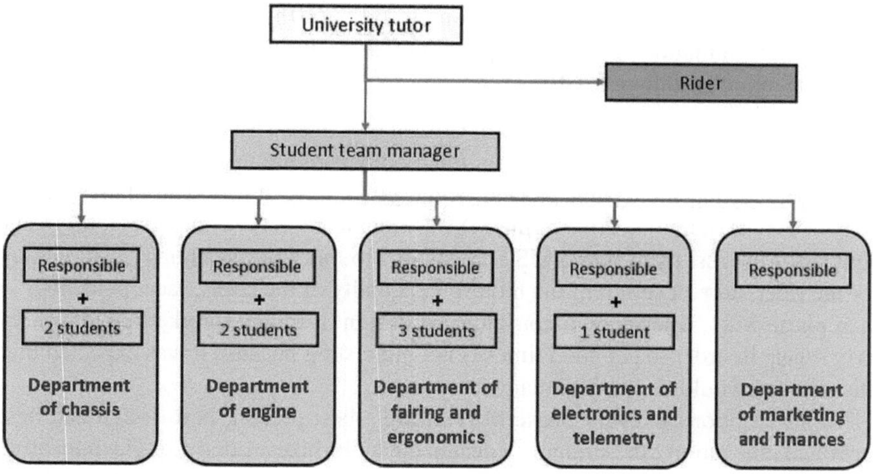

Fig. 1 Hierarquical organization of the MS2-Uniovi team

means of an application where they indicate the position they are interested in and attach their curriculum vitae. Later, all candidates passed an interview to assess their motivation, the adequacy of their curriculum vitae to the position applied for and their knowledge of design, assembly and set-up of motorbike. Rider selection was made by a partner of the team among professional riders of Asturias.

The planning and distribution of tasks was leave on student team manager hands. He had participated in I edition and had influence over the rest of team members because of his age and experience. This student organized the main tasks, assigning a department responsible (head of department), and set a general time schedule for the project.

The responsible of each department should make a detailed planning of their activity and report it on weekly or biweekly meetings. These meetings had official call and a record where comments were collected and new tasks were assigned.

The task of finding sponsors fell on the marketing department, which is also responsible for the promotion and visibility of the team through all possible means (press, radio, TV, web, social networks,...).

The University tutor attended to all weekly meetings and supervised all work done with the heads of each department.

3.2 Project Development

First, and in order to provide students with specific knowledge on methodologies and criteria for design as well as knowledge of advanced use of software, a series of seminars and workshops on the following topics are prepared:

- Conceptual design and proportions of motorbikes (10 h)
- Dynamic simulation and integration with MEF analysis (8 h)
- Research methodologies (6 h).

In addition, a University Extension Course of 4.5 ECTS on advanced design of vehicles was specifically generated for the members of the team.

Before starting the design process, some goals were stated. The phase MS1 is considered the main objective of the MS2-Uniovi team in the competition Motostudent since the results of phase MS2 are subject to more uncertainties (inspired day for the rider, correct set-up of the motorbike, quality of the rivals, race incidents,...) than phase MS1. Therefore, a conventional design of the motorbike that allows an easy set-up in order to get good times is not interesting because it was expected that this strategy would be used by many teams.

To have options in every phase MS1 awards (best project, best design and best technical innovation), the strategy of design the motorbike around a single innovative (for instance, an alternative front suspension) was considered not good. Instead, the innovation in each department is highly promoted, demanding a list of innovations to each department. These innovations were qualitatively and quantitatively assessed, choosing 1 or 2 for each department and rejecting those less technically and/or economically viable.

Then, the specific goals of the motorbike design can be summarized as:

(a) To design a very balanced motorbike from the aesthetic point of view.
(b) To propose an innovative chassis and rear suspension with variable progressiveness.
(c) To incorporate active aerodynamics on the motorbike.
(d) To develop of electronics that allows full sensoring of the motorbike behaviour. In addition, some innovations will be incorporated as, for instance, an electronic mirror.
(e) The development of the engine is limited by the rules of the competition and it is the most expensive part of the motorbike will be not a priority of this project.

Another goal of the industrial project was the establishment of the factory in a country economically thriving, like Brazil, and the implementation of an aggressive marketing campaign.

After conclude the motorbike design, the students started manufacturing some components and purchasing the commercial elements. In this II edition, there was enough money to undertake the purchase of commercial parts and to outsource the specific parts. In addition, there are several machine-tool and teachers and technical staff of the area of engineering of manufacturing processes so that the quality of the chassis parts was high. Students opted to design and manufacture themselves the fairing (including molds) which took a long time but it had a low cost. They also designed, manufactured and calibrated sensors and electronics themselves which allowed them to incorporate innovations such as an electronic mirror.

Students finished the prototype with a delay compared with the planning but in this II edition, they were able to do a circuit test before the final competition.

Table 2 Results for the MS2-Uniovi team in phase MS1 of II edition

Aspect	MS2-Uniovi team
Design and technical calculations	Position 8 of 18
Industrialization systems and cost analysis	Position 6 of 18

Fig. 2 Motorbike and box of the MS2-Uniovi team in Motorland Aragón

3.3 Results in Competition

In phase MS1, the results shown in Table 2 were obtained:

In addition, the MS2-Uniovi team obtained the second prize for Best Design where the jury praised especially the design and manufacturing work done in all aspects of the bike by the team members and their public image. Figure 2 shows the appearance of the motorbike and the box.

In phase MS2, the prototype passed all test on track with the riders of the organization without troubles. In free and qualifying sessions, the MS2-Uniovi reached the 8th position and in race, the team repeated the 8th position.

3.4 Identified Problems

The sequence of scheduled tasks is greatly delayed, particularly in the manufacture. It caused many delays in the assembly and, therefore, the team could only test the motorbike one day in a circuit. Therefore, the student could not learn to do the settings of engine and transmission. It led to delays and errors during the free and qualifying sessions on phase MS2.

The rider did not know the circuit of Motorland Aragon and he rode little the motorbike, not enough to know it properly.

Due to poor timing in the last days before the final, the public presentation of the project was not adequately prepared so the team did not get the expected rating in phase MS1.

4 Conclusions

Benefits for students:

- At knowledge level: additional knowledge to the bachelor in design and testing methods, real experience in racing and business administration, dealing with suppliers and doing a real project.
- At recognition level: 4 ECTS of free configuration, public and social recognition for the prize and contacts with companies in metal-mechanical and automotive sectors.

Benefits for teachers:

- Three research projects (2 IUTA and 1 from University of Oviedo) with funding to investigate the application of active aerodynamics to a motorcycle and to develop a method to design and set up the suspensions.
- Two papers in conferences with the results of the research projects.

Benefits for sponsors and University:

- Print-press appearances: Seven times in regional press and once in national press.
- TV appearances: Seven times in regional TV (TPA and TVE Asturias) in programs of society, news, sports and motor.
- Radio appearances: twice in regional radios (RPA and RNE Asturias).
- Six stands in contest, fairs and exhibitions held in the region.
- Websites and social networks: official team website, Facebook, Twitter, Youtube Channel.

References

1. Henríquez, V., Monzón M.: Impacto de las fórmulas de competición interuniversitaria en la formación académica. In: Libros de actas del XVIII Congreso Nacional de Ingeniería Mecánica, CNIM 2010, Ciudad Real, Poster, pp. 88 (2010)
2. Merino, M.: Motostudent, un paradigma de emprendimiento y educación no formal en la universidad. ArDin, Arte, Diseño e Ingeniería **1**, 63–81 (2012)

Application of a Inter-University Competition on the Enhancement of Engineering Degrees

A. Fernández del Rincón, A. de Juan, P. García, M. Iglesias and F. Viadero

Abstract Within the frame of an innovative teaching project call of the University of Cantabria (Spain), authors have taken advantage of their experience on an inter-university competition (whose main goal is the design and construction of a motorbike among other tasks) in order to apply it on the Mechanical Engineering and Industrial Technologies degrees. The main idea of this experiment is to include a specific activity in each course which were related to this specific competition, using problem and project based learning techniques with the aim of studying some aspects of the analysis and design of components for the motorbike. Two courses, Applied Mechanics and Mechanisms Theory, have participated in this experiment. In the first one, it was proposed a problem about friction of the motorbike's tires, which were modelled as rigid multibody system. In the second one, an open-ended project about the redesign of the rear suspension system of the University of Cantabria original motorbike that took part in the 2010 competition was proposed. Both activities were eligible for students. Results show that the students that took part in the proposed activities pass the courses in a more easy and satisfactory way.

Keywords Project based learning · Problem based learning · Inter-university competitions · Student learning

A. Fernández del Rincón (✉) · A. de Juan · P. García · M. Iglesias and F. Viadero
Department of Structural and Mechanical Engineering, University of Cantabria, Cantabria, Spain
e-mail: fernandra@unican.es

A. de Juan
e-mail: dejuanam@unican.es

P. García
e-mail: garciafp@unican.es

M. Iglesias
e-mail: iglesiassm@unican.es

F. Viadero
e-mail: viaderof@unican.es

J. C. García-Prada and C. Castejón (eds.), *New Trends in Educational Activity in the Field of Mechanism and Machine Theory*, Mechanisms and Machine Science 19, DOI: 10.1007/978-3-319-01836-2_28, © Springer International Publishing Switzerland 2014

1 Introduction

This paper describes an Innovative Teaching Project (ITP) that was carried out by the authors in the academic year 2011–2012. It was based on the experience of the authors in the MotoStudent [1] competition, which is a world inter-university contest where the teams have to design and build a race motorbike prototype. Figure 1 shows the prototype of the University of Cantabria that took part in the first edition of this competition at the MotorLand circuit (Spain).

This experience resulted on an extra motivation for the members of the student's team, who had to face several inconveniences during the process. They applied all the different knowledge they learnt during their degrees, obtaining a global vision of possible future challenges. From the instructors' point of view, the higher achievement was the students' realization that they really solved a problem that initially seemed to be impossible.

On the other hand, this experience had some drawbacks, i.e., the high cost for students (and lack of funds) and the reduced number of students that could take part of it. However, authors liked to transfer the positive learned experience on this topic to the regular courses in the Mechanical Engineering and Industrial Technologies degrees that they instruct. This was done by means of problem [2, 3] and project based learning [4, 5]. These techniques are widely used in learning experiences in engineering all over the world, as can be seen in the literature. Authors also use information technologies as a support of their teaching activity [6].

Fig. 1 University of Cantabria's prototype

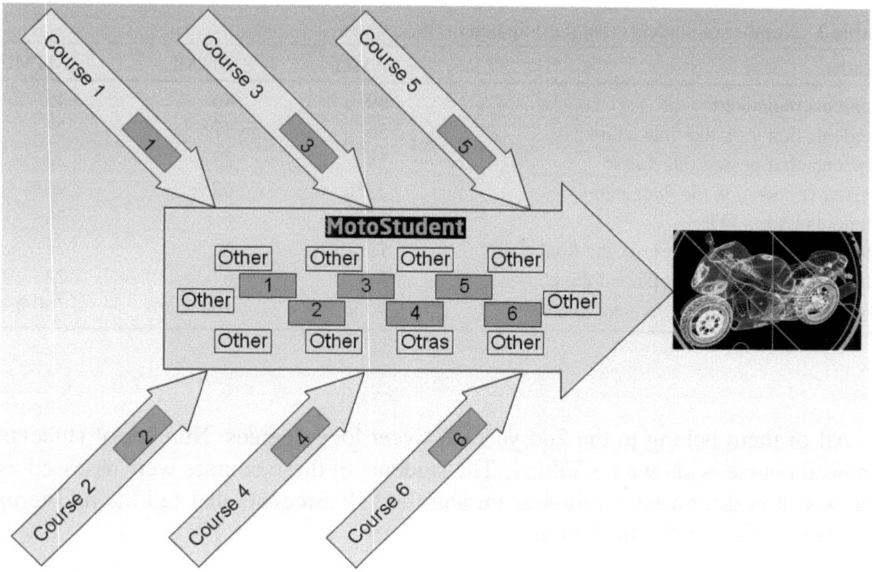

Fig. 2 Main idea of the ITP

2 Method

The main idea of this ITP is described on Fig. 2. It consists on including some activities related to the design a race motorbike within the regular courses of the Mechanical Engineering and Industrial Technologies degrees. The activities are eligible for students and its difficulty and scope is adapted to the level of students' knowledge. However, the activities are problems or projects ill-structured and open-ended, i.e. more realistic problems than those that students are used to deal with in the classroom.

This project aims different goals: (i) to present to students more real-life problems; (ii) to connect different subjects through a common link (inter-university competition); and (iii) to encourage students to take part of different inter-university competitions.

2.1 Participants

Three courses participated in this innovative project, namely two Applied Mechanics (AMME and AMIT) from Mechanical Engineering Degree and Industrial Technologies Engineering Degree respectively and Mechanisms Theory from Mechanical Engineering Degree (MTME).

Table 1 Number of students that participated in the courses

Course	AMIT	AMME	MTME
Students registered	69	46	45
Students that took the final exam	56	37	23
Students that passed the course	18	23	22
Passed versus took the final exam	*32 %*	*62 %*	*96 %*
Students within ITP	15	6	26
Students within ITP that took the final exam	15	6	22
Students within ITP that passed the course	3	6	22
Passed ITP versus took the final exam ITP	*20 %*	*100 %*	*100 %*

All of them belong to the 2nd year of 4 year long degrees. Number of students on each course is shown on Table 1. The students of these courses were assessed as follows: a mid-term test, a final exam and the ITP. Students that fail the mid-term test can retake it in the final exam.

2.2 Procedure

According to plan, the followed procedure to implement the activities into the courses was divided in four main steps. In each one all the agreements were made for the three courses involved in the ITP. A brief summary of each step is described next.

Step 1: Initial coordination meeting among instructors. In this initial meeting, it was agreed to present the same individual problem to the two Applied Mechanics courses (AMME and AMIT). This was because these two courses had 115 students registered in total, and not all of the instructors were available to supervise it.

On the other hand, it was settled that for the MTME course, a project course will be presented and developed by groups. Also, in this meeting it was arranged that all the activities would be eligible.

Step 2: Specific definition of activities within the courses. The ill-conditioned problem for the courses AMME and AMIT was defined in this step. Instructors agreed in not to give to the students all data necessary to solve it, in order to force the students to use information resources. The problem is about a friction issue in the contact between ground and wheel modeled as rigid bodies.

In the MTME course, the redesign of the rear suspension system of the actual motorbike was the train of thought of the course project. Students were given the dimensions of the motorbike and some plans, but they were expected to define even the optimal behavior of the suspension system.

Step 3: Implementation of the activities within the courses. The AMME and AMIT courses are developed in the fall semester of the academic year (from September to February). In order to let students to learn all the necessary concepts for solving the proposed problem, this activity was scheduled before Christmas holidays in one classroom session and after a mid-term test. After that, the statement of work was uploaded to a virtual platform that students usually use during the course. They uploaded their reports in the same way. They consulted their doubts directly to the instructors.

The project course of the MTME was presented to students before Easter holidays, because this course is developed within the spring semester (from February to July). In order to assess the progress of each group, a control session was scheduled at mid semester. Also, an open forum for both students and instructors to give opinions and to exchange information was available in the virtual platform for the course. At the last week, all groups defended in a public session their projects and asked the questions that instructors and other groups asked. They also delivered a report.

Step 4: Assessment of results. The assessment of the problem for AMME and AMIT courses was done directly by the instructors. Students whose report was clearly structured, justified results and with all referenced data obtained the higher marks. This activity graded an extra 10 % of the final mark for the course. This means that a student could have reached a 110 %.

In the MTME, the project course graded up to 20 % of the final mark. This means that a student that did not develop the project only could have a mark up to 80 %. The total 20 % of this activity was divided as follows, 10 % was graded by the report and 10 % by their defense in the final public session. In order to have a feedback of the activity, students were asked to fill in a survey. The students answered questions related to the activity, their learning, the group work and they were also asked to leave comments about their perception of the ITP.

3 Materials

In this section, the materials given to students are briefly described. All materials were uploaded to the virtual platform.

3.1 AMME and AMIT Courses

Statement of work: 'The pilot of a race motorbike is arguing with his team about any kind of irregularity in the contact between the road and the tires. According to the pilot, he had noticed it when he rides the motorbike through the tribune line in a straight trajectory and accelerates, increasing the torque from zero, progressively and linearly. The team realizes that it could be interesting to have a bunch of studies including different conditions. As they have neither time nor resources, they ask to do it to the young and promising students of Applied Mechanics in their second academic year. The team provides students of a simplified rigid body model of the pilot-motorbike set. The suspension system, the air friction and any other frictions

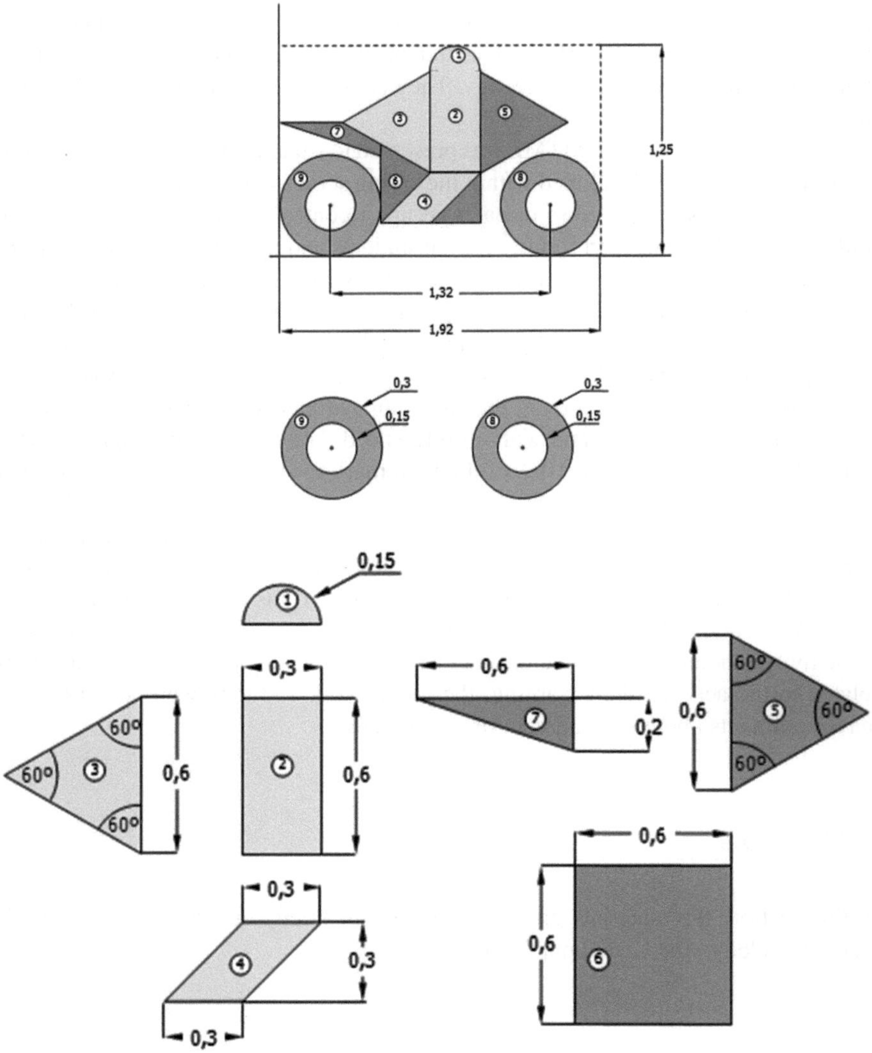

Fig. 3 Additional materials for AMME and AMIT proposed problem

must be not considered. The rest of conditions must be set by the student and they must be clearly described.'

Additional materials: Figure 3 shows the dimensions (in meters) of the simplified rigid body model.

3.2 MTME Course

Statement of work: 'The course project consists on the redesign of the rear suspension system of the University of Cantabria's race motorbike.'

Additional materials: Students were provided by plans of the current suspension system, technical information about the damper, additional bibliography [7, 8] and some tips to develop the final report. As mentioned before, students could use the forum in the virtual platform.

4 Results

Table 1 shows the number of students that were registered in the three courses, the number of students that took the final exam and the number of students that passed each course. Same kind of data is also presented taking into account only students that participated within the ITP.

Results of the survey of the MTME course show that students spent a mean of hours in developing ITP that perfectly matched with the envisaged time foreseen by instructors. They found the activities very useful and motivating and appreciated the instructors work in order to prepare it and solve their queries.

5 Conclusions

As can be seen in the Results section, the ITP has shown positive results among students. Special attention should be put in the number of students that participated in the ITP in the AMIT and AMME courses. Only a few students did participate because the ITP was announced after the mid-term tests. Students obtained very good marks in these tests and the ITP was an extra mark to their qualifications, so they decided it was not worth it. However, in the spring semester for the MTME course, this activity was an important part of the final mark (20 %) and the participation was very much important.

Another aspect deserves to be mentioned about Table 1. AMME and MTME courses belong to the same degree, so students that coursed AMME in the fall semester are the same that in the spring semester coursed MTME. The reader can observe that 23 students passed AMME and are almost exactly the same that took part in the ITP of the MTME course and finally passed it.

In order to improve the ITP in next years, authors would like to highlight some points based on their experience: (i) to encourage students to participate in such activities, it is important that the weighting of the ITP must be at least 20 %. The idea is to increase this weight gradually up to 50 %. (ii) At least two control sessions must be scheduled. The main goal is to solve problems within the groups, for instance, if a member of the group resign to follow the ITP, the group will be at a disadvantage with respect to the others. (iii) The activities must be presented before the mid-term test.

Acknowledgments Authors would like to thank the Vice-Chancellor for teaching staff of the University of Cantabria for promoting the First Call for Innovative Teaching Projects.

References

1. MotoStudent. http://www.motostudent.com/english/index_english.htm. Accessed March 2013
2. Jonassen, D., Strobel, J., Lee, C.B.: Everyday problem solving in engineering: lessons for engineering educators. J. Eng. Educ. **95**, 139–151 (2006)
3. Yadav, A., Subedi, D., Lundeberg, M.A., Bunting, C.F.: Problem-based learning: influence on students' learning in an electrical engineering course. J. Eng. Educ. **100**, 253–280 (2011)
4. Giralt, F., Herrero, J., Grau, F.X., Alabart, J.R., Medir, M.: Two way integration of engineering education through a design project. J. Eng. Educ. **89**, 219–229 (2000)
5. Maseda, F.J., Martija, I., Martija, I.: A training tool and methodology to allow concurrent multidisciplinary experimental projects in engineering education. IEEE Trans. Eng. Educ. **55**, 357–364 (2012)
6. Yilmaz, Ö., Tunçalp, K.: A mixed learning approach in mechatronics education. IEEE Trans. Eng. Educ. **54**, 294–301 (2011)
7. Fernández de Bustos, I., Aguirrebeitia, J., Avilés, R., López, D.: Optimización dimensional de mecanismos incluyendo objetivos de velocidades. In: XVII National Conference of Mechanical Engineering. Ciudad Real (2010)
8. De-Juan, A., Sancibrian, R., Garćia, P., Viadero, F., Iglesias, M., Fernández, A.: Kinematic synthesis for linkages with velocity targets. Systems and manufacturing, J. Adv. Mech. Des. **6**(4), 472–483 (2012)

Making Videos to Engineer Students During the Norwegian Higher Education Reform

E. Villacorta

Abstract This paper is written in order to share the experience of the author recording videos to the students in mechanical engineering at Stord/Haugesund University College in Norway during the Thermodynamic course of 2012. The aim of the author was to create an additional tool to the students and motivate them to learn topics in the course which traditional were considered difficult for a group of students. Nowadays the students, "digital natives", spend a lot of time connecting in internet and grew up around computers at the same time lecturers "digital immigrants" use traditional tools (like books) to share knowledge. The experience of the author, who participated as an online student, in an online course offered by the Massachusetts Institute of Technology (MIT) during spring 2012 is described. During the course, the author got familiar with the different digital tools that were used by lecturers in the online course by MIT, and it was possible to analyze the course design as a lecturer's perspective. The author used a video screen capturing software and recorded three videos to the students who attend the Thermodynamic course in fall 2012. Videos are using intensive in online courses by MIT and the use of videos by other colleagues in Norway. The main idea with the videos was to create a multimedia learning tool that could be used in the virtual learning environment (VLE) supported by internet. During the Thermodynamic course the author tried to measure the student's interaction and perception regarding the videos, this is partial described in the paper. Some aspects regarding the results in the learning approach of the students are described, commented and compared with a previous course during 2011 however this section must be read as the author's experience not as a conclusion of how videos can help to engineer students to get a learning outcome. At the beginning of the academic year of 2012 all the colleges in Norway are involved in a higher education reform as a result of the Bologna Process and at the end of the paper the author share hispersonal

E. Villacorta (✉)
Stord/Haugesund University College, Stord, Norway
e-mail: edmundo.villacorta@hsh.no

J. C. García-Prada and C. Castejón (eds.), *New Trends in Educational Activity in the Field of Mechanism and Machine Theory*, Mechanisms and Machine Science 19,
DOI: 10.1007/978-3-319-01836-2_29,

impression regarding how digital technology can be used by lecturers to meet the learning outcome goals to the engineer students according to the Norwegian higher education reform.

Keywords Blended learning · e-learning · Mechanical engineering · Thermodynamics.

1 Introduction

The author was educated between the 1980s and 1990s during the twentieth century in a developing country (Peru) where he got his professional degree in mechanical and electrical engineering. During a master program in Sweden the author used a Virtual Learning Environment (VLE) for first time. Nowadays, as assistant professor at Stord/Haugesund University College in Norway, the courses content are the same however it is possible to run a class with more tools. The author considers now at least five elements: the reading material (mainly books), the examinations (or compulsory work), the computer, the virtual learning environment and the internet (there are more elements). The virtual learning environment (VLE) is the official channel of communication for the course and in fact the main source of interaction with the students. The digital technology is already at the universities and the author would like to present his experience regarding videos in engineering context.

2 Stord/Haugesund University College and the Norwegian Higher Education Reform

Stord/Haugesund University College (Høgskolen Stord/Haugesund in Norwegian or HSH) showed a lot of interest in developing an arena for e-learning, the description in the paper "Noregsnettet med IT for Open Læring (NITOL)" [1] explains in more detail the accumulated experience in HSH (Since 1990s with Just In Time Open Learning) [2].

At the engineer department we started to work with a new framework to engineer students, and there one skill (among others) which is described like: "Candidate has engineering professional digital competence" [3]. Lecturers must have digital competence thus it is possible to teach and push the students to acquire the digital competence that is required in the new framework. As a lecturer the author started to be active with digital tools during lectures and in the VLE (Fronter). In the case of engineer students the author considers that the challenge is bigger because there are a lot of digital tools and many of them are used in some specific industries and is not good to teach a specific program for example. Another important situation is that in the engineer department we are going to offer a new program with collaboration with

Stavanger University and students are going to take some lectures by video (some of the professors are going to teach from Stavanger).

Many enthusiasts authors (including the author at the beginning) whose celebrate the use of digital technology are very optimist regarding the use of computer in education. Marc Prensky called the young generation "digital native" [4]. But regarding learning and after a pedagogical course the author started to use the word "learning" more close to my learning objectives. However Marshall et al. has a research regarding how engineering student conceptualize learning [5]. It is easy recognize, during a conversation with the students, that many of them are focus in the lower learning level which is called Conception A: Learning as memorizing definitions, equations and procedures. Regarding earning, lecturers would like that students are focus in the Conceptions called C or D (The conception E: Learning as a change as person is a real goal but in the authr's view this goal should be achieved at the end of an engineering program).

3 E-Learning Experience at the MIT

During March of 2012 the author enrolls voluntary in an online course (Circuits and Electronics 6.002x) at Massachusetts Institute of Technology MIT. The course was free of charge and has duration of 14 weeks the evaluation of the course was: "*home works* 15 %, *labs* 15 %, *midterm* 30 %, *and final exam* 40 %. *Each of the home works and labs carries equal weight*" [6]. The midterm exam was during the 8th week and the final exam during the last week.

During the course, the author got familiar with the different digital tools that were used by lecturers in the online course by MIT, and it was possible to analyze the course design as a lecturer's perspective like the evaluation method (portfolio assessment) and the learning activities: electronic tests connected con lectures (a video sequence). After 6 weeks as many online students the author did not continue in the course however there are statistics and a summary of the MIT online pilot project available at the web page [6].

The author interprets this statistics like: 1,54,763 curios, 69,221 potential students, and 26,349 (100 % for the next calculation) students or maybe "curios students" (The author was in this group). Only 7,157 (27 %) achieved deep learning (pass the course). The study hours per week that the author required to earn points in the problems sets, were more than the author expected. The main comment is that the online course at the MIT had a combination of good pedagogical tools (short videos in a sequence) and a lot of students work. It is well known by lecturers that only the student's hard work can make possible to achieve deep learning. Are the teachers at MIT responsible of the remainder 19,192 (73 % students)? Was the course bad? The design of the course was bad? Well for a university like MIT which has a lot of prestige is not possible to admit some possible answers.

The design of the course was clear to the author: push the student to work systematically each week, with some chances during the semester. Is it possible to do that?

At MIT is possible, and as a lecturer the author recommends to find a balance between the study hours and the compulsory learning activities which must be aligned with the learning objectives in a course.

4 Making Videos: Software and the Topics

During the past year 2012 three videos were made to the engineer students in the course of thermodynamics where 90 students were register in the VLE (Fronter). This course is considered difficult for many students and traditional some topics which should be easy for the students became a source of frustration to the lecture. After an informal conversation with a colleague (who has more experience teaching this course) the author realized that the students have difficulties to make a pressure—volume (p–V) diagram or a temperature—entropy (T–s) diagram. In order to give the students an extra tool to fulfill a learning goal, this normally is evaluated both in test (compulsory work) and in exams. The author tried to make short videos but at the end three videos were published and the students would need between 20 or 30 min to watch every video.

The first video: The thermodynamic table video was a first experience to the author. The goal was to integrate many aspects regarding teaching, the book, the allowed aid, etc. It was a test for me and for the students also. The video was published 02/09/2012 in three different links (with three different formats: flash format for computer, Ipad and Ipod or Iphone) after that the question was How could the students react to the video? The author was disappointed about the statistics (students who watch the video), after one week (07/09/2012) only eleven students had watched the video (8 in computer format, 2 in an Ipad and 1 in Ipod or Iphone).

The second video was the p–V diagram and it was publish before a test (08/09/2012). In the test a question regarding the p–V diagram was included and the author expected at many students could answer that question. The written test was held 04/10/2012 and there were 61 students who attend the written test. Before the test only 17 students watched the p–V diagram video, 13 of them attended the written test and the rest (4 students) were not in the test. However the statics regarding the students who watched the video and attend the test (13 students) only 5 students scored pass (pass means at students answer correct over 50 % of the test). Regarding my specific intention to give the students an extra tool and helping to answer a specific question: How to make a p–V diagram? The statistics show a poor result. One student was no able to answer the question, one got half answer and the rest (3 students) were "on the way" to answer the question (between 70 and 80 %). These five students got score between 50 and 58 % in the total test.

The third video was the T–s diagram and it was publish before a second test (18/10/2012). The second test was held on 25/10/2012 and a question regarding the T-s diagram was included but at that time the author was disappointed on digital tools. The second test was an extra opportunity to the students who did not attend and did not pass the first test. However the statics showed that only 6 students watched the

T–s video before the test. There were 33 students who attended the second written test and only 21 pass the test where 3 students watched the T–s video. Regarding the specific goal, the statistics show a poor result again because only 5 students who watched the T-s video attended the second test, two of them did not answer the question and the rest (the three who passed) got a score between 30 and 50 % in that specific question.

There were a third test (the last one) and it was held on 08/11/2012 the author printed out the video attendance 3 days before the test and only 17 students attend the last written test I included a question regarding the T–s diagram again at that point only 9 students had watched the T-s video, and 4 of them attended the third test. One student got 80 % in the specific question regarding the T–s diagram, another student got 30 % and the rest (two students) did not answer the question or got any point.

The author receive at least one specific positive comment regarding the videos however a lot of effort was put in the lecturer side (making videos) and the videos were not watched as the author expected. Deep learning is a very difficult process which involves a lot of time and if the student uses some minutes to watch the video (I am not sure at the students watched the all video or maybe just open it which is enough to be part in the statistics), without any other learning activity, the learning objective is poorly achieved.

The last research about this experience was regarding the student's results in the final exam. It was interesting to know which kind of students watched the video. The author realized that the god students do not need this extra help. Most of the students who watched those three videos were students who got a final grade between 40 and 80 % as a final result in the exam; this is showed in Fig. 1. The discussion regarding quality of the video is not included in this paper, and there were only a written test as main learning activity connected with this topic. During 2011 a colleague who was responsible of the Thermodynamics course gave me the statistics regarding the three tests. Students during 2011 got better results in tests however only a T–s diagram was evaluate in the three written tests and looks like the students at that time achieve a better results regarding this specific topic: How to make a T–s diagram? Probably students in 2011 were more active in that time however as Erstad mention is not good

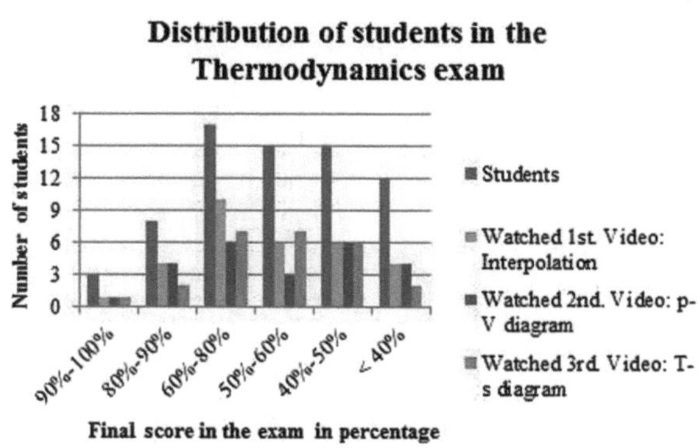

Fig. 1 Distribution of students after exam

to tried to describe a generation of students as "digital natives" without consider the different aspects or categories of media literacy [7].

5 Digital Technologies in the Context of Engineering Education

Regarding learning the author considers that is difficult to make a receipt because every lecturer designs his course with specific learning objectives and some learning activities which must be aligned with the learning goals in the course [8]. In my experience as an e-learner student at MIT [6], during 2012, I had the opportunity to use videos an deal with electronic tests during the course. And in the author's opinion the electronic test is underestimated as a learning activity. In the following description I tried to remark the differences from the learning perspective the advantage of an electronic test:

The written test, which normally used at HSH, has many advantages and in the engineer department is used as compulsory work. The main advantage is that students are prepared to the exam and they take the test in the same conditions that will be evaluated during the exam. However in many occasions some students return the answers before the test ends (they think at they did right) and the students have to wait (at least two weeks) for a feedback. In that period of time the learning activity, design by the teacher, could disappear. Another aspect is that the teacher only receive the answer notes, it is not possible to know how students think, only what they had written. The time between the students work with the question and get a feedback is long and during that time the students can lost the original learning intention of the teacher. This is shown in the Fig. 2 where the "two weeks arrow" makes the feedback not important in terms of learning:

The electronic test (which was underestimated by the author and is nearly the same that the written test) works different and it is possible to use in both ways, not only to

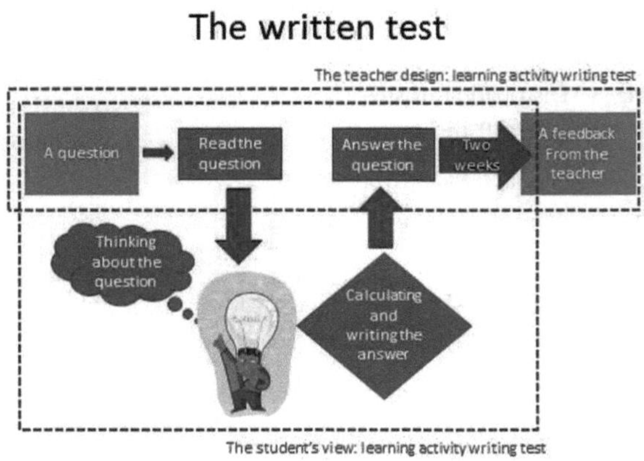

Fig. 2 The written test: The Lecturer design and the students view

measure if the students achieve the learning goal in addition the lecturer can get good information regarding the students learning progress. The author changed his opinion regarding electronic test after the e-learning course in MIT. The homework's were really electronic test and students had to achieve the right answer thus the computer gave a student an immediate feedback and if the answer was wrong the student had to calculate again (check errors during calculation) or think again about the question in order to get the right solution. During this process (to do it and do it again if it is necessary) a student, who gets the right answer, comes to a deep learning zone this is described in Fig. 3. What about the rest? What about the students who are not able to get the right answer? In the MIT course was not possible to survive without pass, but here at HSH the lecturer is able to take the statistics of an electronic test and use some minutes in the next class to help to the students who did not achieve the learning objective thus the electronic test can be a good tool for both lecturer and students.

This paper is written to share the author's experience not as a conclusion of how videos can help to engineer students to get a learning outcome. Despite the poor results the author considers that short videos (MIT used videos between 7 and 9 minutes in sequence) combined with obligatory electronic test could work in an engineer context, as is presented in Fig. 4. Some topics requires an extra help after a lecture because is difficult that all the students focus 100 % of attention during 45 min lecture. In the next figure the author shows how the digital tools can be combined every week (or chapter) during the semester.

However the author has a lot of concerns regarding digital tools and maybe some concerns will be part in future work. N. Carr has a book called "The Shallows, What the Internet is Doing to our Brains" in the book he presented some neuroscientist

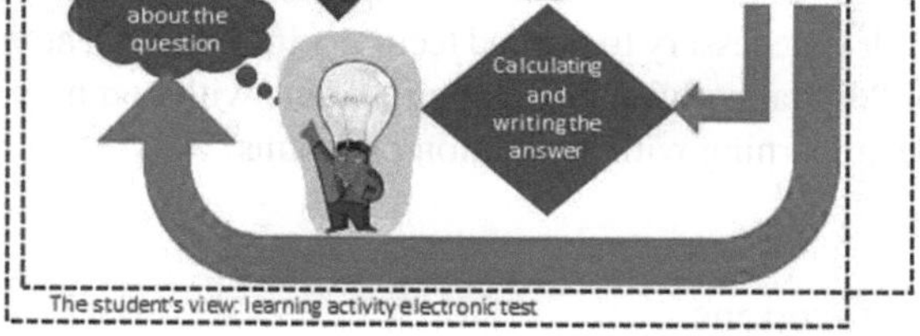

Fig. 3 The electronic test, the Lecturer design and the students view

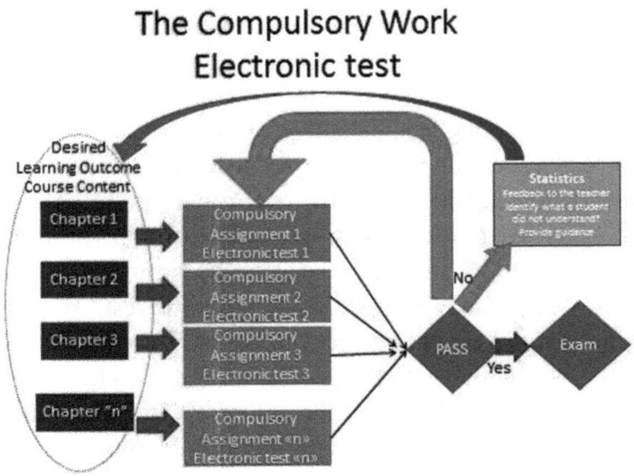

Fig. 4 Systematic compulsory work with electronic tests

papers and argued that internet is changing us in many aspects [9]. One part of his discussion is regarding ethics and the author thinks that we have not discussed this enough.

6 Conclusions

The author emphasizes that this experienced needs to be evaluated with more tools thus a proper conclusion section must not be presented. The author considers that as engineer educators, we must discuss about the impact of digital technology in learning activities at universities:

The author understands that technologies like: VLE, e-learning has many advantages and these technologies are parted of our present and future development, but are we sure that the long term results will be positive? Are we sure that we can achieve deep learning with this tools?

The statistics in the MIT online course shows that just a few online students got a diploma (achieved deep learning). What are going to do the others universities? Follow the new online trend despite of the poor learning results?

It is necessary to pushed lecturers into a digital arena? It is easy to pushed students or "digital natives" in the digital arena with complex software? It is easy to achieve deep learning with simulation programs?

References

1. Ask, B: Noregsnettet med IT for Open Læring (NITOL) - et pionerprosjekt innan nettbasert læring. nubok, Norway (2009)
2. Krumsvik, R.J.: Digital competence in Norwegian teacher education and schools. High. Educ. **1**(1), 39–51 (2011)

3. Universitets- og Høgskolerådet, Nasjonale retningslinjer for ingeniørutdanning, (2011)
4. Prensky, M.: Don't Bother Me mom—I'm Learning. Paragon House, US (2006)
5. Marshall, D., Summers M., Woolnough, B.: Student's conceptions of learning in an engineering context. High. Educ. **38**3, 291–309 (1999)
6. Massachusetts Institute of Technology: Circuits and Electronics 6.002x. www.mitx.mit.edu (2012)
7. Erstad, O.: Educating the digital generation exploring media literacy for the twenty first century. Nordic J. Digit. Lit. **5**(1), 56–71 (2010)
8. Biggs, J.: Enhancing teaching trough constructive alignment. High. Educ. **32**, 347–364(1996)
9. Carr, N.: The Shallows, What the Internet is doing to our brains. W.W. Norton and Company, New York (2011)

Mechanism Theory in Architecture Education

M. Barej, S. Hoffmann, G. Della Puppa, M. Hüsing, B. Corves and M. Trautz

Abstract In mechanical engineering tasks concerning motion and moving parts are very common, thus of course topics like kinematics and mechanism theory are content of teaching in various lectures, exercises and laboratories. At RWTH Aachen University besides mandatory basic courses for all engineering students, continuative courses for students specialized on engineering design or automotive engineering impart profound knowledge. In architecture education the focus lies on the design of buildings in consideration of social, functional, esthetical and statical aspects. Although due to the demand of sustainable and adapting buildings the importance of deployable structures and kinematic parts is increasing, these topics are rarely found in education. Suitable opportunities to bring interested prospective architects in touch with mechanism theory are design projects scheduled for master students. Following an existing cooperation in the research field of foldable structures the Chair of Structures and Structural Design (Faculty of Architecture) and the Department of Mechanism Theory and Dynamics of Machines (Faculty of Mechanical Engineering) offered in the winter semester 2012/13 an interdisciplinary task of

M. Barej (✉) · M. Hüsing · B. Corves
Department of Mechanism Theory and Dynamics of Machines, RWTH Aachen University, Aachen, Germany
e-mail: barej@igm.rwth-aachen.de

M. Hüsing
e-mail: huesing@igm.rwth-aachen.de

B. Corves
e-mail: corves@igm.rwth-aachen.de

S. Hoffmann · G. Della Puppa · M. Trautz
Chair of Structures and Structural Design, RWTH Aachen University, Aachen, Germany
e-mail: hoffmann@trako.arch.rwth-aachen.de

G. Della Puppa
e-mail: gdella_puppa@trako.arch.rwth-aachen.de

M. Trautz
e-mail: trautz@trako.arch.rwth-aachen.de

J. C. García-Prada and C. Castejón (eds.), *New Trends in Educational Activity in the Field of Mechanism and Machine Theory*, Mechanisms and Machine Science 19,
DOI: 10.1007/978-3-319-01836-2_30,

designing a foldable bridge to a group of architecture students. The focus of this student project was on the design of a new foldable bridge, but in order to enable the students to solve this unusual task some preliminary activities were performed. The project started with analyses and presentations of existing solutions followed by lectures and workshops providing contents from architecture, arts and mechanism theory. Observations of the project showed that thanks to this preparation students considered kinematic issues as well as architectural ones from the very beginning. The aim of this paper is to present how mechanism theory can be integrated into Interdisciplinary education, which content the authors consider being important for students and which influence can be observed in students' results.

Keywords Interdisciplinarity in education · Architectural design · Student project · Movable structures

1 Introduction

Tasks and challenges for mechanical engineers often contain the design and development of kinematical parts and mechanisms. Therefore students of mechanical engineering are prepared for solving these motion related tasks from the very beginning of their studies. This preparation is provided by lessons and exercises considering among other issues kinematics, dynamics as well as the choice and calculation of machine elements like screws, bearings or drive belts.

Since in architecture moving parts are not that common, here issues concerning motion play a subordinate role in education. Today besides singular functional demands also global aims like adaptability or sustainability force architects more and more to deal with deployable structures. Challenges set by moving parts in architectural scale have to be faced by interdisciplinary teams, but this cooperation requires some preparation and previous knowledge.

In order to bring this into architecture education the authors offered the task of designing a deployable bridge as design project for master students of architecture. This paper presents this interdisciplinary educational project, the requested and the offered support, the students' outcomes and the results of an evaluation performed afterwards.

2 Design Project 'Rheinhafenbrücke'

A design project typically is the main semester task of an architecture student during the master course [1]. Project skills, abilities, methods and individual understanding of design are taught during the course of the project. Thereby designing is conceived as the integrality of planning, arranging, constructing and reflecting.

Fig. 1 Model of Existing Bridges: **a** Diffené-Bridge in Mannheim reconstructed by H. Wen **b** 'Buckelbrücke' in Duisburg reconstructed by S. Hussain

The task of a design project consists of two subtasks, a seminar (3 CP) and the project (15 CP). The seminar is thematically leant against the project task and is meant to give the individual student deeper background knowledge in the issue of the particular field of action, usually by the analysis of built examples. The project consists of developing a design which has to take functional, architectural and urbanistic aspects into account. At project end the students have to present their results in kind of drawings, models, visualizations by photos and renderings and, if necessary by textual explanations. After each presentation the lecturers ask questions and give critical suggestions. The grades given to the students are based on fulfillment of the given task by the student's design, on the degree of detail but also on the quality of the presented models and posters.

The project 'Rheinhafenbrücke' was proposed at RWTH Aachen University in the winter term 2012/13. Within this project a transformable pedestrian bridge had to be designed. It was supposed to have a span length of 60 m and a width of approximately 2.4 m.

The students could select out of two proposed building sites, the 'Medienhafen' in Düsseldorf and the culture island in the 'Phönixsee' in Dortmund. Due to the complexity of the task it was not mandatory but recommended that the students work in teams of two persons. The design project was supervised weekly by the scientific assistants, architects and engineers, of the Chair of Structures and Structural Design. Moreover, three intermediate colloquia were held with all participants and all tutors, in which the work items were discussed and advanced. Furthermore once a week a seminar with the topic 'Transformable Structures' was performed. Here students mainly analyzed examples of transformable bridges, e.g. [2, 3], but also dealt with Origami topics. They were asked to build movable models of both Origami and architecture examples (Fig. 1). By this way an intuitive greater understanding of transformable structures was obtained. At the end of the semester, the students presented their projects in a final exam colloquium. Some of the outcomes are shown in Chap. 4.

3 Complementary Teaching Activities

Since the proposed task was unusual for architecture students, complementary teaching activities were necessary. The educational concept was supplemented by optional lessons and workshops as well as additional mentoring in order to provide supporting content. These offers pursued three different objectives: inspiration, qualification and supervision.

Inspiration is necessary to help the students in designing innovative motion concepts. Therefore examples from different application fields and from the Japanese art of paper folding Origami have been provided. As a guest lecturer the Origami-scientist Tomohiro Tachi offered a talk about 'Computational Design of Origami Form' including a lot of folding patterns and visualizations. Further he held a workshop concerning 'Rigid Origami Structures', in which he presented some of his models and software tools [4] for the simulation and analysis of rigid Origami [5]. Afterwards the students had the opportunity to try out the software as well as manual paper folding (Fig. 2a).

After the students have developed their own idea or vision of a deployable bridge it was time to enhance their qualification. Since this project was situated in the master program it was estimated that the students were familiar with basics of architecture and structural engineering. Therefore topics typical for mechanical engineering and mechanism theory were required. These issues were provided in two presentations titled 'Best of Mechanism Theory' and 'Desired and Undesired Motion'. Two presentations cannot be sufficient to impart all necessary knowledge. Especially since the students designs were strongly different causing different questions and requiring different assistance. The aim of these presentations was to give hints about expected problems and approaches for related solutions. Initially different types of mechanisms and joints, application fields for mechanism theory and the design process for mechanisms were introduced. Those students who still were looking for inspiration and required more examples of existing mechanisms were relegated to appropriate literature and the 'Digital Mechanism and Gear Library' [6]. Furthermore methods for analysis as well as for structural and dimensional synthesis were presented [7]. Here the focus lied on planar mechanisms and graphical methods. To enable the students to model and analyze their kinematical concepts intuitively at an early design phase 'Cinderella' [8] was proposed as a helpful tool (Fig. 2b). Besides contents related to

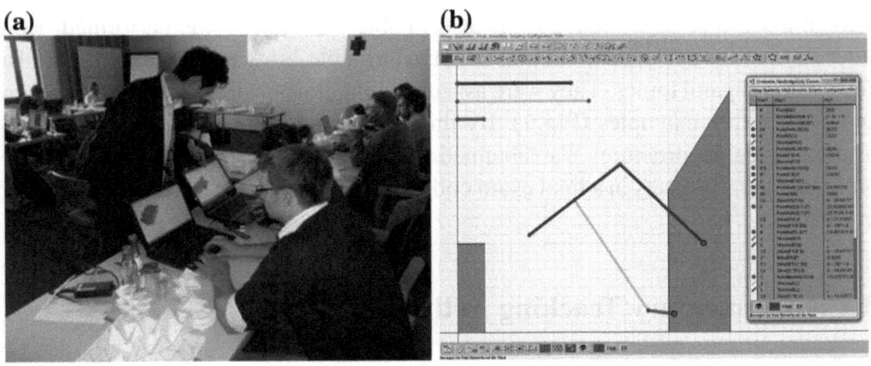

Fig. 2 Complementary Teaching Activities: **a** Origami-Workshop with T. Tachi, **b** Screenshot of a Construction in Cinderella

the design and concept phase also first hints concerning the technical realization of joints or bearings were given. In the second lecture the focus was on contents related to machine dynamics. Topics like actuation concepts, motion design, occurring forces or vibrations were treated. Although these aspects were beyond the students' task their mention sensibilized the participants for further technical problems.

The third objective: supervision was achieved by the attendance of a mechanical engineer within the regular sessions and colloquia. In this way the students had the opportunity to receive feedback and discuss specific problems concerning their kinematic concepts.

4 Student's Outcomes

Inspired by the analysis of existing bridges as well as by the mechanisms and Origami-based structures presented within the complementary teaching activities the students' designs follow very different concepts. Not only the appearances but also the complexities of developed structures differ strongly, as depicted in Fig. 3.

Among the solutions there are designs based on planar mechanisms with degrees of freedom equal to one (Fig. 3d) or two (Fig. 3a). In some concepts the reduction of actuating forces by use of counterbalances is intended (Fig. 3c, e). Other designs use serial structures and require sequential (Fig. 3b) or combined actuation (Fig. 3g). In one of the concepts even a spatial mechanism based on Origami is applied (Fig. 3f). Further differentiating factors are the choice of electric motors or hydraulic cylinders as actuators and the transmission of actuating forces by rigid links or by cables.

5 Evaluation

Since this is the first time such an interdisciplinary educational project was offered to students the evaluation is essential for the preparation of following projects in coming semesters. Therefore on the one hand students have been asked to review the project in complete but also the optional offers themselves. On the other hand based on the student's results the lecturers reviewed which contents had the expected benefit and if further content is necessary.

5.1 Student's View

Nine out of fourteen students agreed to participate in a voluntary and anonymous evaluation, which was carried out before the final exam colloquium. This point in time was chosen with the intention that the grades do not influence the rating.

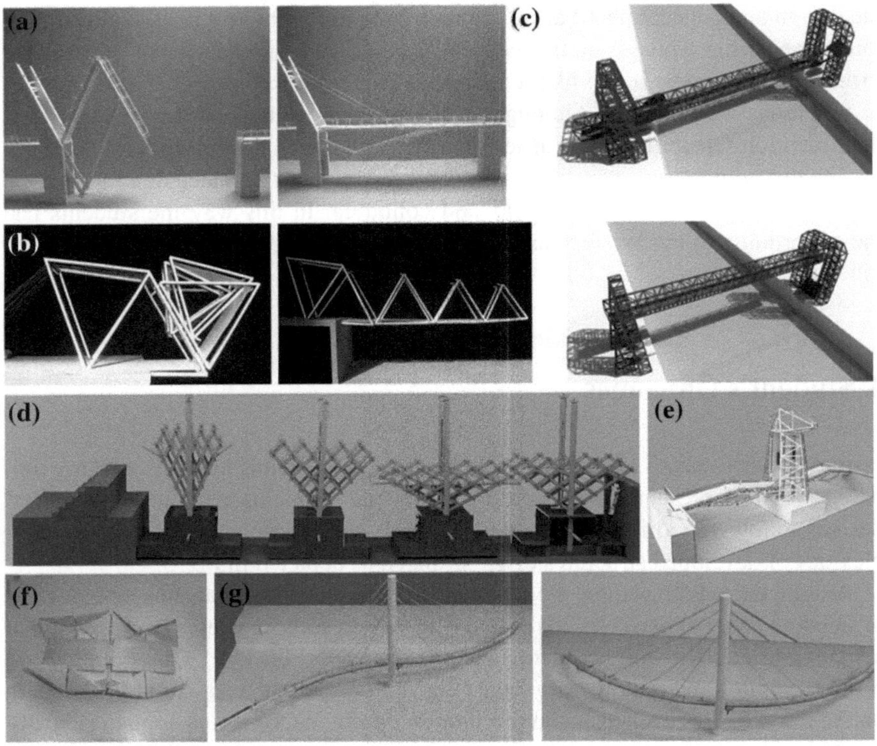

Fig. 3 Students' Outcomes: **a** A. Gibas, **b** J. A. García Poveda, **c** I. Llana García, A. Martin Barreiro, **d** L. Merk, H. Wen, **e** A. Leoni, K. Longhi, **f** M. Schwab, **g** D. Sauerbrey, C. Schröder

A questionnaire was handed out, consisting of multiple choice questions and space for own suggestions.

Table 1 contains essential results of the evaluation. The first three courses were open to a greater audience and were held beyond the schedule of the design project. That might be the reason why these additional offers were less frequented. The fact that only one student decided to base his design project on Origami-principles may explain the stated relatively low relevance of these topics for the students' concepts. The two events 'Lecture about Mechanism Theory' and 'Colloquia' took place during the normal timetable and only the participants of the project 'Rheinhafenbrücke' were allowed to attend. In both, the lecture and the colloquia, they had the opportunity to ask specific questions concerning their own designs. This might be an explication for the good grades given by the students.

Table 2 contains the students' suggestions for improvements. From this prospect it can be derived that the participating students request more practical, constructive assistance. From students' view the amount of information about abstract, artistically inspiring topics was sufficient.

Table 1 Results of evaluation by students (5—very strong / 1—very weak)

Event	Participants	Relevance
Lecture about origami	5 / 9	Ø 3,2
Origami workshop	4 / 9	Ø 2,5
Lecture about motion	4 / 9	Ø 3,8
Lecture about mechanism theory	7 / 9	Ø 4,6
Colloquia	9 / 9	Ø 4,3

Table 2 Students' suggestions for improvements (5—very strong / 1—very weak)

Event	Relevance
More practical examples (Bridges)	Ø 4,1
More origami	Ø 2,6
More other transformable constructions	Ø 3,9
More structural synthesis	Ø 3,9
More dimensional synthesis	Ø 3,8
More construction details	Ø 4,5

It can be stated that the most relevant outcomes of the evaluation are the need for compliance of the additional offers with the schedule and the demand for more technical supervision.

5.2 *Lecturer's View*

The overall level of the works showed on the one hand the promising potential of kinematical issues in architecture students' projects; on the other hand it suggests directions of further development for a future continuation of the activity. With the final stand of their works, the participants have proved to be able to assume reviews from external technical areas as from mechanical engineering and to implement suggestions for improvement. The extension of the didactical content to a wider explanation of machinery elements is supposed to bring an enhancement in students' design choice.

Furthermore the great contribution in terms of creativity and spontaneity that was brought in by the students should be integrated from the early stage of the activity in a more synthetic view, translating the concept design into simple functional schemes of mechanism. To achieve this didactical goal, a more intensive presence of a mechanical engineering tutor at early stages of supervision sessions is likely to be profitable.

6 Conclusions

The presented concept for an interdisciplinary inspired design project in architecture education gives students the opportunity to deal with contents which are typically beyond the topics of their studies. The project's outcomes, the student's dedication and their feed-back show the high interest for this type of projects. Based on the positive experience gained within this project it is desirable to continue this cooperation.

From the mechanical engineer's point of view movable structures in architecture build a fascinating field of application for mechanism theory. For this reason it is to examine how to integrate students of mechanical engineering into the next design project. This would give students of both faculties the chance to collaborate continuously in an interdisciplinary team and thereby to enhance their knowledge as well as their social skills.

However the great interest aroused among the students on such topics represents a positive feedback on their behalf and suggests a positive outlook for the establishment of a continuative collaboration between the institutes in didactical activities, alongside the already experimented common research path.

Acknowledgments The authors want to thank T. Tachi for the inspiration provided by his lecture and workshop as well as the Exploratory Research Space at RWTH Aachen University for making his visit possible.

References

1. RWTH Aachen University: Description of examination regulations: M.Sc. in, Architecture (2011)
2. Freudenberg, G.: Zwei neue Waagebalken-Klappbrücken in Mannheim. In: Stahlbau 2/1989, pp. 35–43 (1989)
3. Bühler, D.: Brückenbau, Deutsche Verlags-Anstalt GmbH (2004)
4. http://www.tsg.ne.jp/TT/software/, Accessed 13 Mar 2013
5. Tachi T.: Simulation of rigid origami. In Origami: Peters A.K. (eds.) Proceedings of 4OSME, pp. 175–187 (2009)
6. http://www.dmg-lib.org/dmglib/main/portal.jsp?context_setLangCode=en, Accessed 13 Mar 2013
7. Kerle, H., Corves, B., Hüsing, M.: Getriebetechnik. Vieweg+Teubner (2011)
8. http://www.cinderella.de/tiki-index.php, Accessed 14 Mar 2013

Interdisciplinary Factory Planning in the Education of Mechanical Engineers and Architects

A. Kampker, M. Trautz, S. Kamp and I. Leufgens

Abstract The current and future challenges in the field of factory planning result from social, political and economical developments and exceed solely technical advances of production engineering. An interdisciplinary approach that achieves results in the fields of social sciences as well as engineering is inevitable to respond to future challenges in factory planning. Only by merging the special knowledge of different disciplines, the field of factory planning will be prepared for future needs. In order to pursue this interdisciplinary approach, a multidisciplinary research group was founded at RWTH Aachen University. The Chair of Production Management (Faculty of Mechanical Engineering) and the Chair of Structures and Structural Design (Faculty of Architecture) at the RWTH Aachen University are part of this research group. Besides research activities, one goal of the research group is to encourage an interdisciplinary education in the field of factory planning in which production technology and building technology are integrated. Up to now, the cooperation in teaching was implemented in three projects in the topics of Automotive Industry, Solar Energy and Machine Tools offered for master students at the faculty of architecture. In the course of these projects, teachers of both chairs gave lessons and guided students of architecture in the process of planning industrial buildings. The interdisciplinary formation of teachers and students fosters learning from each other's expert knowledge and prepares students for future tasks in interdisciplinary planning teams. The abilities to work in a team and to find solutions in a creative dialogue are encouraged in real or fictional projects. The different levels of knowledge

A. Kampker · S. Kamp (✉)
Chair of Production Management at the Laboratory for Machine Tools and Production Engineering, Faculty of Mechanical Engineering, RWTH Aachen University, Aachen, Germany
e-mail: S.Kamp@wzl.rwth-aachen.de

M. Trautz · I. Leufgens
Chair of Structures and Structural Design, Faculty of Architecture, RWTH Aachen University, Aachen, Germany

I. Leufgens
e-mail: ileufgens@trako.arch.rwth-aachen.de

J. C. García-Prada and C. Castejón (eds.), *New Trends in Educational Activity in the Field of Mechanism and Machine Theory*, Mechanisms and Machine Science 19,
DOI: 10.1007/978-3-319-01836-2_31, © Springer International Publishing Switzerland 2014

of the students and the various methods in the field of factory planning pose particular challenges. The essential knowledge is imparted by a collective lecture to facilitate the communication between the disciplines. At the end of the course, each project is the result of different disciplines whose exigencies have been synchronized.

Keywords Factory planning · Interdisciplinary teaching · Architecture · Production engineering

1 Introduction

The manufacturing industry has to deal with social and economic changes. The factory requirements become more and more complex and uncertain. A key factor for competitiveness is to deal with this complexity and these uncertainties [1]. Changing preconditions and a turbulent economic environment call for two innovations in factory planning. First, due to shorter product life cycles companies do accept less and less mistakes, rework or cost and overrun during factory planning projects. As a result, more efficient and interdisciplinary planning processes have to be mastered. Second, complex social and economic factory requirements demand an interdisciplinary approach, which combines all disciplines that are involved in a factory planning project. The involved disciplines range from socio science, architecture to mechanical engineering and further [2, 3].

The upcoming factory planner should have the skills to work in such interdisciplinary teams. The skills cover several aspects. A fundamental requirement is the ability to work in a team. Furthermore, a basic understanding about the other planning tasks is needed and the technical language has to be understood. If the planner does not understand the specific names and sayings, communication will be harder and information will be lost. All participants have to know the interfaces between their tasks and all other disciplines. Most important, every planner has to be aware of the impact of his own planning result on other tasks and the overall project targets.

Today, there are no interdisciplinary courses at university. Neither the faculty of mechanical engineering nor the faculty of architecture or another department offers interdisciplinary courses. Only similar fields of study, such as mechanical engineering and industrial engineering or architecture and structural engineering, are working together. This is due to the organizational structure of a university with its faculties. The faculties are organized by disciplines and most courses are held within a faculty.

An interdisciplinary education can only be reached, if courses are offered by entities of different faculties. Interdisciplinary skills can not only be taught by lectures. It has to be learned by experience. Therefore an interdisciplinary course should consist of a theoretical lecture as well as a project work, which has to be worked on in a team. Such courses can be introduced in two steps. First step is to offer a course, which is opened for one field of study and is supervised and coached by an interdisciplinary team of tutors. Second step is to open the course for different fields of study, e.g. architecture, structural engineering and production engineering, so that interdisciplinary teams work on projects and are supervised and coached by an interdisciplinary team of tutors.

2 Factory Planning in the Education of Architects

The joint research activity of the Chair of Production Management and the Chair of Structures and Structural Design in the field of factory planning is incorporated since winter term 2010/2011 by conjointly supervised courses. The aim is to encourage the understanding of students for the processes within a plant and for the complexity of an interdisciplinary planning situation.

During the course, the students had to develop an efficient framework for changeable plants in the topics of Automotive Industry, Solar Energy and Machine Tools. Special attention was paid to the urban integration, the requirements of the production process, the internal organization and the corporate architecture. The building should offer space for production, administration, technical and media support and representative space for events. The design of each area corresponded to the respective requirements. It was requested to design an innovative and efficient structure and façade for the production with a strong emphasis on scalability and changeability of the building. The presentation of the brand requires a certain transparency that allows visitors to participate in the production. Equally important is the design of the factory according to the desired image of the manufacturer. Exemplary results from the previous years will be presented in the following passages.

2.1 Factory for the Production of Electric Vehicles

The project focuses on the production of the electric vehicle "Street Scooter", a project of the Chair of Production Management. Since the novel and innovative technology of e-mobility is subject to continuous change, the development of versatile buildings is one main aspect of the task. The aim of the project is to develop an innovative concept that handles both, claims of versatility and flexibility as well as user requirements, constructional and technical aspects.

The draft of Inga Hausmann B.Sc. is a compact plant structure which is implemented in a vertically integrated module. The addition of the module creates a bolt, which allows the necessary flexibility of the factory by modular expandability. The factory is positioned on the western edge of the industrial park Avantis on the German-Dutch border in direct relation to the motorway A76. The completed product is already visible from the highway: through showcases in the façade in which the final "Street Scooter" is stored and exhibited. The factory bolt is divided into three areas. On the side facing away from the highway, the entrance area and the visiting area are positioned. The visitor is able to walk through the factory via stairs and lifts to see the various manufacturing steps in the production without disturbing the operational procedures. Next to the visiting area, the column-free production is arranged on several levels. The sequence of the individual production steps takes place in the building from the bottom upwards. This keeps the heavy machinery, like presses and equipment for painting on the lower level of the factory. In each level offices and technical rooms can be hooked and supply cores are positioned, flexible and directly related to the required production area. On the side of the highway is a shelf for delivery, storage and transportation of materials needed between levels as well as for the technology and staircases.

In total, a highly organized structure is emerged, which allows a high degree of communication between the different areas by its vertical orientation and represents an innovative approach to the design of industrial buildings (Figs. 1, 2 and 3).

2.2 Factory for the Manufacture of Solar Modules

The content of the project is to design a factory for the manufacture of solar modules. As the landscape extends the south of the property to the North Eifel, the factory should be as far as possible integrated into the landscape and use the sloping terrain to the south. The hall for the production is built in the hillside and is not visible from the street. Only the tower, which includes the administration of the company, stands out as a landmark. The aim of the draft by Hendrik Goossens B.Sc., is to

Fig. 1 Exterior view

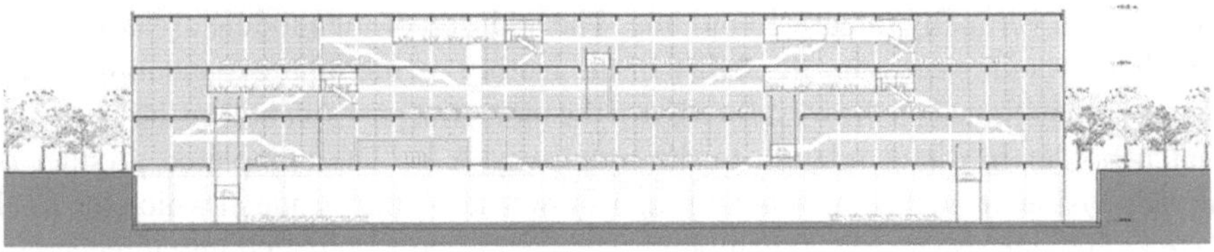

Fig. 2 Longitudinal section through the production levels

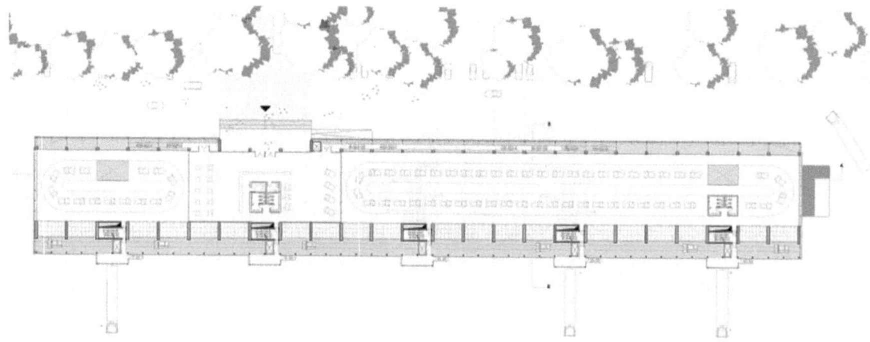

Fig. 3 Floor plan of the entrance area and production

Fig. 4 Section of terrain and factory

make the environmental advantages of the product visible in the production progress and the company's headquarter. The roof of the hall is greened intensively and this greening is only interrupted by the skylights and the tower. The skylights allow visitors to walk on the roof and follow the production and increase the natural light level in the production area. The exposure to light is also done via the glazed south facade. By burying the production hall, the north wall and the soil form an optimal thermal mass. The factory consists of a vertically organized tower and a horizontally organized production hall, which intersect in the bottom part. The tower is located centrally in the factory between storage and production and is accessed from the north side on the greened roof of the hall. The tower divides the production hall in two areas, the storage in the west and the production in the east. The incoming and

Fig. 5 Perspective illustrating the walkable roof

Fig. 6 Exterior view of the factory

outgoing materials are transported through two large doors to the factory loading dock. On the supporting wall and the south facade, the transport of the material required for the production lines and the carriage of the finished solar modules to the storage takes place. The production area offers space for two production lines and the quality control (Figs. 4 and 5).

2.3 Factory for the Production of Machine Tools

The production building by Stefano Pascale B.Sc. and Tobias Anne-Lehre B.Sc. in the cross-border business park Avantis meets the complex requirements of the production as well as the increased demands for a versatile architectural factory building and the reasonable integration into the landscape by its spatial layout and the whole shape of the overall building shape. The building is a hall, with an area of

Fig. 7 Perspective of the manufacturing area

240 m × 52 m. The floor plan is column-free and allows an extension of the building along the longitudinal axis. The demands on the span were met with a supporting structure of prefabricated reinforced concrete frames as a folding construction. The interior of the hall is designed as a large room in which different functional areas can be set in. From this approach, the entire administrative area of the factory, together with catering and changing facilities is located as a "house within a house"—solution in the production hall. The equality of employees and the high quality of workspace is similar to the assembly and production areas: the offices are lined up along the eastern facade and receive an overall view of the landscape. The folding of the roof allows the natural lighting of the hall, emphasizes the opening into the surrounding landscape of Aachen and gives the building its distinctive architectural character (Figs. 6 and 7).

3 Conclusion

Due to the positive feedback of the participating students and the need for an even more interdisciplinary course, a course offered by different disciplines e.g. mechanical engineering, architecture, civil engineering, energy engineering and industrial engineering is planed. The course "interdisciplinary factory planning" addresses students of these fields. The students have to plan a factory based on information about the product, the processes, the strategic goals and the actual production site respectively the green field area. The given information is real data to ensure a realistic case.

The project contains value stream design, capacity planning, material an information flow planning, layout planning, structure planning, technical building equipment planning, investment planning et cetera. By applying knowledge across subjects, the skills of each discipline are imparted in an interdisciplinary teamwork as part of this course. In the future, the course will be offered by different faculties to the master degree courses in mechanical engineering, architecture, civil engineering, energy engineering and industrial engineering.

The main purpose of the course is to show the existing interfaces between disciplines to the students and to plan the specific aspects simultaneously, e.g. production system, building and technical equipment. Coming along with simultaneous planning, project management is been taught.

The course will start with a two-day lecture. During the lecture the students are introduced to the basics of each disciplines. The case data will come from a partner company. The project work starts after the lecture. During the project work research assistants of the participating chairs coach the students. Up to three intermediate colloquia are held. The students have to show their intermediate results and get feedback. The final draft will be presented at a colloquium at the end of the course.

4 Figures

1, 2, 3 Lehrstuhl für Tragkonstruktionen: Fabrik zur Herstellung von Elektrofahrzeugen von Inga Hausmann, 2011.

4, 5 Lehrstuhl für Tragkonstruktionen: Fabrik zur Herstellung von Solarmodulen von Hendrik Goossens, 2011.

6, 7 Lehrstuhl für Tragkonstruktionen: Fabrik zur Herstellung von Werkzeugmaschinen von Stefano Pascale und Tobias Anne-Lehre, 2012.

Acknowledgments The authors want to thank the RWTH Aachen University for funding interdisciplinary teaching. Furthermore the authors want to thank the students Inga Hausmann B.Sc., Hendrik Goossens B.Sc., Stefano Pascale B.Sc. and Tobias Anne-Lehre B.Sc. for publishing their drafts.

References

1. Sardar, Z.: Welcome to postnormal times. Futures **42**, 435–444 (2010)
2. Kampker, A., Osebold, R., Trautz, M.: Innovative Fabriken interdisziplinär planen. In: wt Werkstatttechnik, online (102), No. 4 (2012)
3. Wiendahl, H.-P., Reichardt, J., Nyhuis, P.: Handbuch Fabrikplanung, Hanser Verlag, Munich (2009)

Knowledge Pills for Teaching of Mechanism and Machine Theory

C. Casqueiro, R. Maceiras and A. Suárez

Abstract From the beginning of its activity, simultaneously with the implementation of European Higher Education Area (EHEA), the Defense University Center located in Spanish Naval College has developed a great effort to improve the learning methods. This effort takes two main directions: improve of classroom activities and the implementation of e-learning systems. These systems are a way of support for the students attending our classes, facilitating subsequent study. One method used for e-learning are the knowledge pills. However, although the pills are widely used for theoretical explanations, at Mechanical Engineering studies the explanation of basic exercises and the problems resolution provides a new and more difficult scene. Teaching of Mechanism and Machine Theory is a good example of these cases. At 2011–2012, we developed a knowledge pills collection for this course, like a guide for the resolution of kinematics problems at planar mechanisms. The student can follow a sequence of learning, from the resolution of position to the accelerations problem, passing for the velocities. In this work, the authors aim to show the learning sequence chosen, the characteristics of created pills concerning at their duration and structure, and the tools used for save them and edit.

C. Casqueiro (✉) · R. Maceiras · A. Suárez
Defense University Center, Naval Academy attached to the University of Vigo, Pontevedra, Spain
e-mail: ccasqueiro@cud.uvigo.es

R. Maceiras
e-mail: rmaceiras@cud.uvigo.es

A. Suárez
e-mail: andres.suarez@cud.uvigo.es

J. C. García-Prada and C. Castejón (eds.), *New Trends in Educational Activity in the Field of Mechanism and Machine Theory*, Mechanisms and Machine Science 19,
DOI: 10.1007/978-3-319-01836-2_32,

1 Introduction

The implementation of European Higher Education Area (EHEA) has required an important effort for improving the teaching and learning process [1]. In this work, the implementation of an e-learning tool for the Mechanism and Machine Theory subject is presented.

The Defense University Center located in Spanish Naval College began its activity at October 2010, with the first course of Mechanical Engineering. Nowadays, the third course is running. The Mechanism and Machine Theory is a subject of second course, and was implemented last year.

Since the beginning of the course, subject teachers together with other colleagues from the center tried to offer students the means to facilitate independent learning. Implementation guidelines of EHEA expressly include autonomous work of the student as a part of your learning process. The particular conditions of the students of the Spanish Naval College, which combine engineering studies at the university center with military training, makes it even more necessary to strengthen the capacity of independent learning. In addition, for students who travel and live away from home is appropriate that this learning can occur at any time and place.

In this case, the knowledge pills were chosen as an appropriate multimedia means to facilitate independent learning [3], and the knowledge pills collection developed at 2011–2012 course, like a guide for the resolution of kinematics problems at planar mechanisms, forms part of the means provided for the e-learning of our students of Mechanism and Machine Theory [2]. This work explains the characteristics of these learning material.

2 Knowledge Pills in CUD-ENM

Knowledge pills are a tool that is being used recently. This type of resource is useful in the e-learning processes, exclusively for distance teaching or to support classroom teaching. The knowledge pills are often called learning objects [4].

These multimedia objects are a short video (usually 7–15 min) composed by two images: the teacher image and the document image. The document can be a Power Point file or another equivalent text or graphic resource. The presence of the speaker provides enhanced communication capability, because the use of the body language. This way, using gestures allows better transmit the emphasis on the points of greatest interest.

Usually, the knowledge pills must transmit an only one concept. This way, the video can have a limited time and the student can keep attention. However, some knowledge pills are longer, because they develop a concept of relative complexity. Often it is the case of exercises, because is needed an introduction for remember the theoretical concept before the resolution, and this must be sequential for a good compression by the student.

Use of knowledge pills allows to student a non presential learning. In our case, the student combines the classroom with the autonomous work, and the knowledge pills are oriented to reinforce your non presential work. It is the case of knowledge pills that have been developed.

2.1 Technical Characteristics of Knowledge Pills

The of knowledge pills employed at CUD in the Theory of Machine and Mechanism are a video composed for two images. The building image has 1280 × 720 pixels, correspondent with HD video format, like other knowledge pills by University of Vigo [4]. In these videos, the teacher's image is combined with an image obtained by a Power Point file. This document shows schematic texts and images. The lecturer explains the file content as explains with whiteboard, but the camera only record the teacher image (Fig. 1).

The study of record has two screens. One screen is placed in front of lecturer, and can show the Power Point content or information like an autocue. The other screen is placed right of lecturer, and shows the Power Point content. This screen is used for the teacher can signal over it the content that needs more emphasis. When this video is combined with the Power Point file, the area dedicated for this is placed at right of teacher, too.

One difficulty for the teacher is unable to bring the text or figures with a laser pointer. It is necessary to use other techniques to highlight the contents. The teacher

Fig. 1 Teacher in front of camera for recording

Fig. 2 Situation of the teacher and the two screens

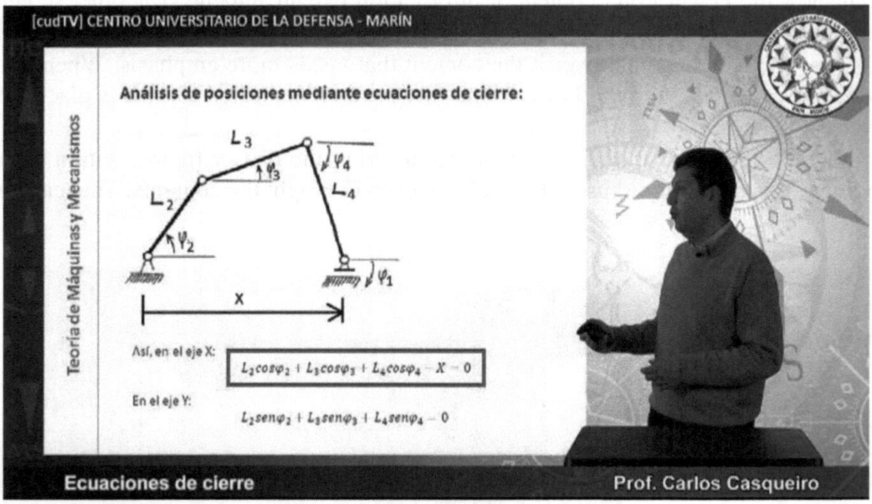

Fig. 3 Capture of knowledge pill in HD video

can look to whiteboard and after processing the image seems the teacher is in front, like a classroom, but their actions are limited (Fig. 2).

After the record, the two captured images in 4:3 format [5] must be combined with specific software for the video manager for generating a HD video, 1280 × 720 pixels (Fig. 3).

2.2 *Storing and Serving the Pills*

In the Defense University Center located in Spanish Naval College, there is a website that acts as repository of videos, called cudTV (http://tv.cud.uvigo.es). This website publishes educational pills created by teachers or researches of this university.

Published content in cudTV is generated by PuMuKIT tool, developed by University of Vigo. This is an audiovisual content manager dedicated to published multimedia objects, like a videotaped class or conference. The multimedia objects stored, knowledge pills or another resource, are arranged into series. One of these series can group, for example, the documents corresponding with a same subject, the sessions of a congress... [6].

In addition to accessing the different series by direct way, PuMuKIT allows create a list of documents recommended, for example by their date of creation. Using the capacities that PuMuKIT provides, the website cudTV offers a list of contents grouped in series described by a name and a brief explanation, through the section called "Mediateca" and offers the possibility to access the section "Novedades" ("new"), which includes the latest documents added to the database (Fig. 4).

The series employed in CudTv groups the different knowledge pills as a function of its subject. We currently have seven groups, six for teaching (mathematics, chemistry, English, new technologies, mechanics, graphical expression) and one dedicated to various issues relating to the operation of the center. Within each group, the videos are separated by subject. So, in the section dedicated to the mechanics have a set of knowledge pills for theory of machines and mechanisms, strength of materials, materials engineering and thermodynamics and heat transfer. The courses taught in the CUD, for mechanical engineering, with varied subjects make the contents of these are diverse and educational needs are different. Some subjects used mostly text. Others need support graphics. Many of these require the use of mathematical

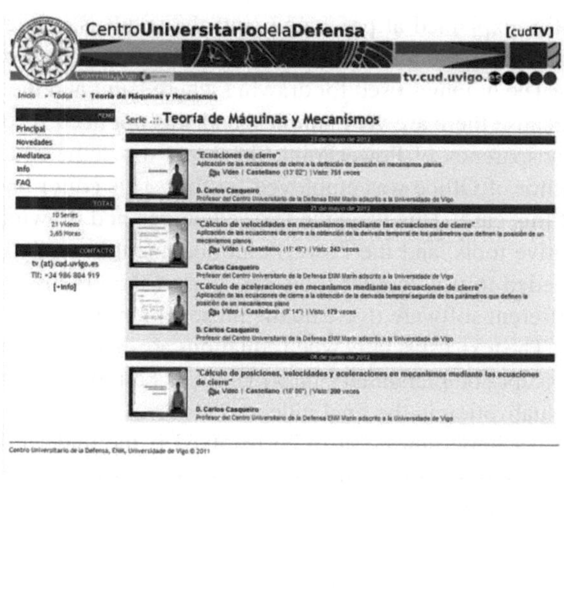

Fig. 4 Screen capture with cudTV website

formulas. Addition, some subjects focus their production on theoretical pills while others need treat numerical problem solving.

3 Knowledge Pills of Theory of Machine and Mechanism

Most knowledge pills show theoretical concept, but the knowledge pills developed for this subject are focused in the exercises resolution. The series is structured in four knowledge pills (Fig. 4). The first shows the position analytic resolution mechanisms. The second knowledge pill shows the calculations of speed by the Jacobian to the equations obtained before. The third shows the resolution of accelerations, for a complex know of kinematics of mechanism. In the three learning objects, the theoretical explanation is reduced to minimum, and the method is explained by basic samples. The last knowledge pill resumes the process applied to a sample.

At this series, the complete method is explained of sequential way. But each resolution needs the previous step and this is remember in the knowledge pill a brief way. Then, the calculation of accelerations is explained beginning by obtaining the position equations after applying the Jacobian for the velocities and finally, explains the method for obtain the accelerations. The collection is completed with a fourth knowledge pill about the application of the method in a more difficult sample. This way, the student learns the complete method by parts, in a simple case, and finally he acquires the capabilities for solve a more real case.

The duration of these knowledge pills is between 9 and 16 min. The longer one is a little excessive, but it is corresponding by the last video that explains the method for the resolution in a complex mechanism. It is a singular case, because the student must have yet the knowledge and this video helps him for understanding a resolution of an exercise viewed in the classroom.

Technically, the knowledge pills employed for this subject follows the characteristics explained at previous point: the result is a video in 1280 × 720 pixels (HD video) and the teacher appears at right side (for the observer).

The graphics used for drawing where Paint and the PowerPoint drawing utilities, because there are very simples, and was not need software more complete. Most of transparency of PowerPoint file shows too some equations (Fig. 5). The equations editor of Office was employed for this. This editor has limitations, but it is enough at this case. This way, the text, equations and drawings can be created with Office native tools, and the PowerPoint document is easily edited. In other cases, can be needed to use another tools and import the graphs or equations later, or employ a different software to create the presentation file.

The text, especially when equations are displayed, coud have a 24 points minimum (except complementary information as bottom of page). Minor size can have a bad visualization in the final video.

An important point to be resolved is the underlining of the main data, display information sequentially, and shows how each data is introduced into the equations. In the authors' previous experiences other methods were used, but in this case the

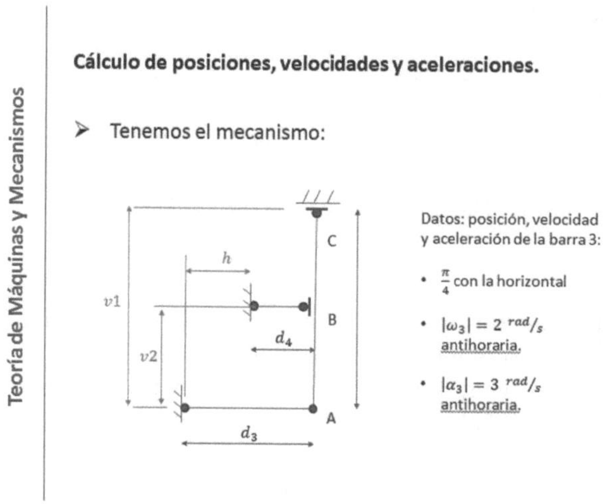

Fig. 5 Sample of PowerPoint slide with graphic and equations generated with native tools

selected method was show in sequential way on the screen the previously created information, use different colours for highlights and surround the parts what must be featured with lines, circles and other forms. This way requires more time for editing that the use of manual tools over touch screens or tablets as iPad used in some cases [7], but is more precise. The subject requires accurate and clean drawings. The mathematical expressions with differentials and other symbols must be written with correction for a good understand, too. For this reason, the manual method for drawing and writing was not used although it is faster than the complete computing edition.

4 Results and Discussion

The principal result of this work is a collection of knowledge pills that develop the kinematic study of mechanism with an analytical method. This collection was structured for being overall used or separately in function of student needs.

The videos can be used in a PC screen for better experience but the resolution is adequate for see in a tablet of medium or great format (usually 8″ or more).

The collection was created and offered to students later than the correspondent classroom, but before the exam that contains this theme. The acceptation among the students was very good. The next table shows the number of visits for each video (Table 1).

This results show that knowledge pill with fewer viewing has a 20 % of visits more than number of students after two courses. The knowledge pill with highest number of viewing has a 492 % of visits more than students. Although resources can

Table 1 Number of visits per knowledge pill

Theme	Duration	Visits
Positions	13′ 02″	748
Velocities	11′ 45″	239
Accelerations	9′ 14″	179
Sample of application	16′ 06″	197

be seen from the outside, website cudTV is not famous in the university world today, and most of the visits are of our students.

It is interesting to appreciate the lowest number of visits of the third knowledge pill, that explains the calculation of accelerations, very easy when was understood the method for calculate the positions and velocities. The final sample of application has a bigger number of visits than this knowledge pill. The more successful knowledge pill was the first, that shows the principals concepts of method.

Regarding to the obtained qualifications, it can be pointed out that an improvement of about 20 % was observed in the final exam comparing the part of exam dedicated to this issue (before the creation of the pills) with the corresponding part of final exam (when the pills were available). We can not directly attribute this improvement in the qualifications to the more successful knowledge pills, but it should be noted that these were the only educational material added after the first examination.

5 Conclusions

After creating this collection of knowledge pills and use them during the past course and also in the present, it can be concluded:

- Series of knowledge pills help students in their learning process, and improve their understanding.
- Perform knowledge pills to explain concepts and practical exercises requires a suitable method for structuring content. These should be displayed sequentially.
- The duration of knowledge pills is relevant for the successful. Too long duration of video reduces the viewings number.
- Manual methods to mark the content and improve the vision of the highlights can not be applied in some cases. The precision of the graphics and the correct display of some mathematical symbols are impaired.

Moreover, from the use of knowledge pills can be established new challenges in its construction. For future work, it would be interesting to find a more dynamic method of creating graphics without losing accuracy seems interesting.

References

1. Felder, R.M., Brent, R.: How to improve teaching quality? Qual. Manag. J. **6**(2), 9–21 (1999)
2. Jiang, P.P., Tong, G.Y., Jiang, C.X.: Analysis disadvantage of multimedia education. In: Proceeding of the 2007 First International Symposium on Information Technology and Applications in Education, ISITAE, 23–25 Nov 2007, pp. 456–457
3. Sanchez, A., Maceiras, R., Urréjola, S., Cancela, A.: A new learning tool: solving exercises in a knowledge pill. In: INTED2010 Proceedings, pp. 5561–5566 (2010)
4. Wiley, D.: Learning objects: a definition. In: Kovalchick, A., Dawson, K. (eds.) Educational technology: an encyclopedia. ABC-CLIO, Santa Barbara (2002)
5. Sanchez, A., Cancela, A., Maceiras, R., Urréjola S.: The use of learning pills combined with flashcards on chemical engineering teaching. In: INTED2010 Proceedings, pp. 454–458 (2010)
6. Goyanes, V., Casar, A., Pousada, J.M., Gonzalez, R.: IPTV for educational multimedia content distribution. Uvigo-TV, ARCA and PuMuKIT projects in Spain. EUNIS, http://eunis.dk/papers/p69.pdf (2008)
7. Ulloa Sande, C., Rey, G.D., Alfonsín, V., Sánchez, A.: Use of i-pad applications in the design and production of a collection of pills for solving heat transfer problems. In: EDULEARN12 Proceedings, pp. 1459–1464 (2012)

Evolution of the Supervised Work in Machine and Mechanism Theory at ETSEIB

S. Cardona Foix, L. Jordi Nebot, R. Pàmies-Vilà and P. Català Calderón

Abstract This paper presents the evolution of a supervised work carried out in small groups of students included within the compulsory subject Theory of Machines and Mechanisms at Barcelona School of Industrial Engineering—ETSEIB. The implementation of the supervised work started during the academic year 1999–2000 and it has remained till the academic year 2011–2012 with the idea of being an opportunity for the students to put in practice specific knowledge and general skills. The paper explains the experience and enhancements done by the teaching staff during all these years in order to fulfill with the expected objectives related to the supervised work.

Keywords Supervised work · Large groups · Real mechanisms work · Evaluation

1 Introduction

Theory of Machines and Mechanisms is a compulsory subject set on the fourth semester of bachelor's degree in Industrial Technology. Previously to Bologna process, this subject was known as Machine Theory in the previous Industrial Engineering degree. Teaching environment, objectives and contents have remained the same for both subjects. Theory of Machines and Mechanisms has SIX ECTS teaching

S. Cardona Foix · L. Jordi Nebot (✉) · R. Pàmies-Vilà · P. Català Calderón
Department of Mechanical Engineering, Universitat Politècnica de Catalunya,
ETSEIB, Barcelona, Spain
e-mail: lluisa.jordi@upc.edu

S. Cardona Foix
e-mail: salvador.cardona@upc.edu

R. Pàmies-Vilà
e-mail: rosa.pamies@upc.edu

P. Català Calderón
e-mail: pau.catala@upc.edu

J. C. García-Prada and C. Castejón (eds.), *New Trends in Educational Activity in the Field of Mechanism and Machine Theory*, Mechanisms and Machine Science 19,
DOI: 10.1007/978-3-319-01836-2_33,

load: 5 credits are taught in a classroom with 75 students—following the classical structure of conventional lectures and resolution of 2D exercises inspired mainly in real mechanisms. The remaining credits are taught in laboratory sessions with 25 students.

Despite the difficulty of being a subject with large number of participants—as an average 230 enrolled students—during the academic year 1999–2000 a supervised work was introduced. The supervised work was thought as a good opportunity for the students to practice the specific knowledge acquired during the subject, knowledge of previous subjects and for developing general skills.

The supervised work consists in the realization of a kinematic and dynamic analysis of a real mechanism proposed by the students. The proposed objectives related to the work are: (i) develop a certain culture for recognizing mechanical elements or basic mechanical groups in real examples and situations; (ii) use computer programs for the calculation and simulation of mechanisms; (iii) learn, by means of rigid body theory, the basic and operative tools for developing kinematic, static or dynamic analyses in mechanisms and machines; (iv) write a technical report and present it orally; (v) practice team work skills.

During all these years, hundreds of mechanisms proposed by the students which can be studied with two-dimensional models have been gathered. Figure 1 shows an example of a mechanism and the simplified model proposed by the students in order to analyze it by means of Mechanism and Machine Theory.

Since its first edition, the supervised work has been realized in groups of three students and has had a weight of 20 % in the final mark of the subject. The first edition was the only one where the supervised work was set as an optional choice instead of a laboratory exam. From the academic year 2000–2001, the supervised work has been mandatory and has replaced the mentioned exam. In order to achieve the expected objectives in a better way, during the academic year 2002–2003, three tutored sessions dedicated to the realization of the supervised work were established. This way of acting was followed till the last semester of the academic year 2010–2011, where

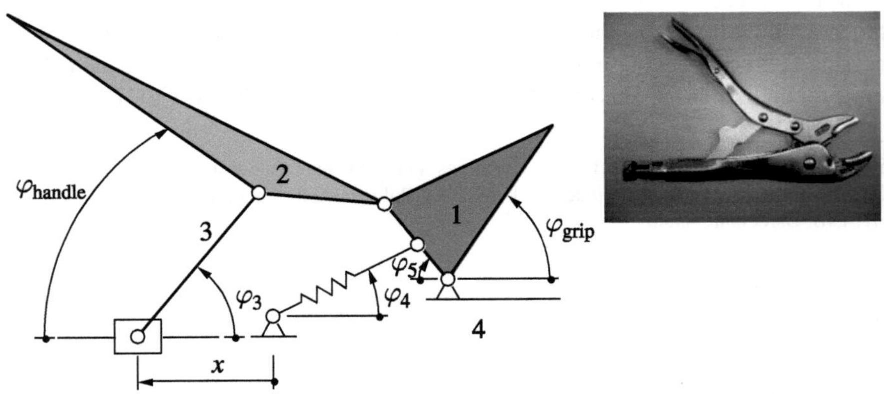

Fig. 1 Universal grip pliers. Real mechanism and a proposed model

a large increase on the number of the students was noticed. On the academic year 2011–2012 a new methodology was established in order to face the increase of teachers' load without losing expectations and maintaining the same objectives.

The purpose of the authors is to share their experience and all the changes implemented—fixing a tutor, development of an own program of mechanism simulation, elaboration of pattern documents, introduction of the peer-to-peer assessment, creation of rubrics—when supervising works in the scope of compulsory subjects with large number of students.

2 First Methodology (2002–2011)

In order to achieve with the objectives of the supervised work, soon it was noticed that specific sessions for its realization were necessary. Three tutored sessions were established.

On the first tutored session, the supervised work and its objectives were presented and in a period of 2 weeks the students should accomplish with the following tasks:

- Choose an adequate mechanism for being simulated with the simulation program learned on a previous laboratory session.
- Draw to scale and with standard symbols, with a CAD program, the schematic mechanism diagram and propose an appropriate set of generalized coordinates to describe the system like in Fig. 1.
- Draw to scale with a CAD program, the schematic diagram dimensions for each solid of the mechanism.

These tasks were materialized in two deliveries that had a weight of the 15 % of the supervised work. The students should refresh the usage of a CAD program learned in previous subjects. The students had plenty of time to search for an appropriate mechanism and if a mechanism was not found, the tutor could offer one from the Machine Laboratory.

On the second tutored session, the deliveries were given to the teacher who realized a general revision in front of all the groups—presentation, width of lines or schematization of solids. An appointment with the tutor was fixed if students wished a more accurate revision. The next tasks were:

- Fix a set of interesting objectives to study the mechanism. In general, two kinematics and two dynamics objectives were required.
- Implement the rigid body system into the simulation program and make kinematic and dynamic analyses.
- Present the obtained results in a set of graphics like in Fig. 2.
- Revise, if it was necessary, the previous schematic diagrams of the mechanism, deleting the mistakes and introducing the required issues for study the fixed objectives.

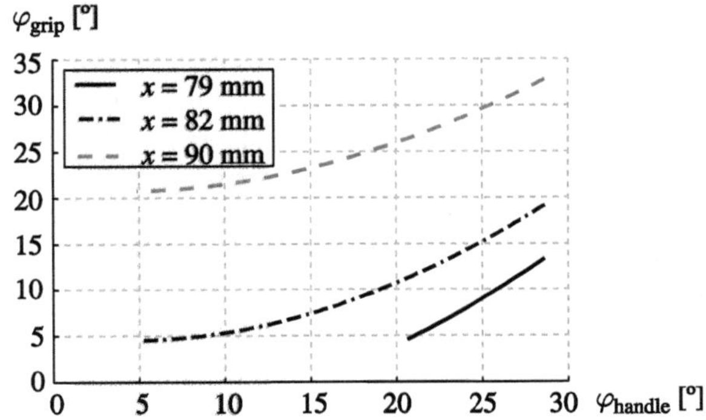

Fig. 2 Example of a kinematic result presented in a graphical form for the universal grip pliers

These tasks were materialized in deliveries that also had a weight of the 15 % of the supervised work. The main objectives of these tasks were to acquaint students with the use of computer programs for the simulation of mechanisms, obtaining results and presenting them in a graphical form.

At the beginning the simulation program DynaPack was used, but during the academic year 2002–2003 an own program simulation—PAM: Programa d'Anàlisi de Mecanismes [1]—was developed as a simulation tool. The usage of existing commercial package of mechanism simulation was avoided because a large learning process is required. PAM was thought as an easy tool for introducing students to the simulation programs and being able of learning basic operation with a laboratory session of 1.5 h. In order to smooth the progress of students' learning, during the academic year 2006–2007 a self-study course with the animation software Macromedia Flash was elaborated [1].

On the third tutored session, general corrections of the deliveries were done in front of all the students. All groups were asked to make a quick explanation of the mechanism studied, simplifications done and fixed objectives. Then, tutor explained the content that must contain the final report and the oral presentation. For the development of the final report, the document entitled *Technical and scientific report presentation* [2], was developed for the teachers' staff in accordance with the UNE 50135:1996 (ISO 5966:1982).

Finally, on the last week of the semester, students performed the oral presentation in front of a committee formed by two teachers. It must be pointed out that this presentation is one of the first done during their degree.

These three tutored sessions were fixed in order to guide the students along the work and so they had an idea which tasks should be accomplished. The major part of the tutored sessions, the tutor explained the tasks and little work was done during the sessions. This way of proceeding required several appointments outside the tutored sessions, especially when the mechanism had to be implemented on the simulation program.

With this methodology, good results were achieved and the average of passed works had been of 94.5 %. During all these years, each tutor, each semester had been supervising an average of 15 works (approximately 45 students).

However, during the spring semester of the academic year 2010–2011, which was the penultimate semester of the previous Industrial Engineering degree, the teaching load was largely increased because the enrolled students passed from an average of 230 students to 350, and the number of teachers was the same.

3 Second Methodology (2011–2012)

A new methodology with more tutored sessions and different evaluation method was implemented. The tutored sessions passed from 3 to 6—including the final presentation. The main changes are explained below.

In a previous laboratory session—before the first tutored session—students were asked to organize themselves in groups. Out of the classes and in groups of three or four students, they had to search some plausible planar mechanism that could be studied using the simulation program. During the first tutored session or even before, students and teachers agreed on the final selected mechanism and in the same session the groups could make the schematic mechanism diagram by hand. This change allowed the tutor to see students working in class and detect mistakes that, with the previous methodology, were detected on the second session. Moreover this way of acting facilitates afterwards the drawing with a CAD program. The tasks for the second session and the weight—15 %— were the same as the previous methodology.

On the second tutored session, again started with a general revision of the schematic mechanism, but it was dedicated to implement the system to the simulation program. The objective of the session was to achieve the movement of the mechanism and, for that, both the geometry and also the kinematic pairs should be defined correctly.

The third session was dedicated to obtain the first results. Students were allowed to see the mechanism in motion and moreover they implemented the dynamic magnitudes of their system. After that, the two kinematic and the two dynamic objectives of the work were discussed. It was seen that, with the mechanism in movement, it was easier for the students fixing the objectives of the supervised work. The tasks for the fourth session and their weight in the mark were the same as in second tutored session of the previous methodology.

At the fourth session the students gave the graphical results that represented the kinematic and dynamic objectives. Again, the delivery was generally corrected to return feedback to students. After that, teacher explained what the final writing report had to include. The task for the next session was to develop a first report that was evaluated for the classmates. In this case, the final report extension was fixed in six pages and a document with what should contain the writing report was developed [3].

At the fifth session, using a rubric, classmates evaluated both their own writing report and the one presented for the other groups. The rubrics were done in a way that the students evaluated different aspects of the writing report and gave a mark between 0 and 10 [3]. Also, a pattern document was prepared where students, a part from the mark, should explain why they gave this punctuation. So at the end of the session, the marks were given anonymously to each group but also the reason why they had given this mark. The mark given by the classmates had a weight of 5 % of the final work's mark. Note that the rubric used was published at the beginning of the semester in the digital campus, so the students knew how they were evaluated. The rubrics and the change of giving a second version of writing report helped to avoid typically errors and proved as a good tool because the students were critics. At the end of the session, the teacher explained how the oral presentation should be faced.

On the last session, oral presentation was done in front of a committee formed by two teachers and also in front of the rest of the classmates which had to evaluate with another rubric [3].

With this methodology, when supervising the work, good results were achieved and the average of passed works had reached 100 %. The elimination of the failed works is done thanks to do peer-to-peer assessment process. Also with this methodology, required appointments outside the classroom were significantly reduced.

4 Discussion of the Experienced Methodologies

Comparing both methodologies, teachers' staff conclude that with the first methodology where the students had 15 days for choosing a mechanism, the sort of mechanism were more original. Also, because the tutored sessions were lower and they had no view of the classmates' works, the works were more different in the extension but also in the content. Furthermore, with the first methodology requires of more appointments outside class for solving problems, especially close to the delivering date.

On the other hand, with the second methodology with more tutored sessions, the students rely more on the guide process and the creativity in the mechanisms chosen and in the originality of the content is reduced. This drawback is accepted due to the subject is set on the first stages of the degree (fourth semester) and it is on future subjects like bachelor's final project where the students can develop the originality and creativity skills. Directly asked to the students with an open question-answer Quiz the students where happy with the implemented peer-assessment process—although some feels that empathy could deviate the classmates' mark—and with the opportunity of enhance the final written report. The peer-assessment process has proof as an effective tool to detect errors on time and it helps to explain the elimination of failed works.

Regarding also the peer-assessment process, on Fig. 3 it is compared the final mark given by the teachers—in red—and the classmates' mark—in blue—of the supervised works. Analyzing the results of Fig. 3 it is concluded that for the students

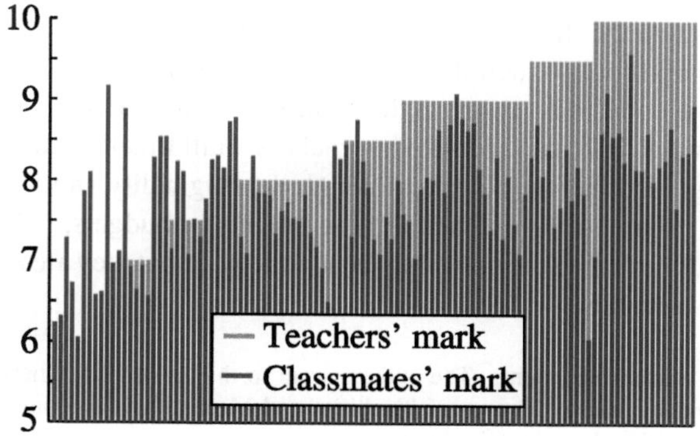

Fig. 3 Teachers' mark and classmates' mark

is hard to give good marks higher than 8.5 but also it is hard for them give lower marks (below 7) in comparison with the teachers' marks.

Also during all these years a lot of different mechanisms have been chosen for the students. The most popular mechanisms have been the garage door, elliptical bicycles, strength machines, tools like universal grip pliers, steering direction mechanism of radio control cars or scissor jack. But also there are some curios mechanisms like Theo Jansen's strabdbeest, antique toys, automata and mechanical games or walking dolls.

Sadly for the teachers' staff, due to the cutback on the university budgets, this way of acting has been changed for the nowadays academic year due to the teachers' staff has been reduced. Right now the 20 % weight of the supervised work has been split. A 10 % corresponds a laboratory exam and the other 10 % corresponds to a simulation exercise based on Artobolevski's linkage mechanisms [4].

5 Conclusions

In this paper two methodologies have been presented for supervising works in subjects with large number of students experienced for the Machine and Mechanism teachers' staff from the academic year 1999–2000 till the academic year 2011–2012. The experience is considered successful both from the students' side and the teachers' side. From the students' side because high level of passing works have been reached in both methodologies and for settling basics concepts of how to write technical reports for other subjects. From the teachers' side the methodology allows achieving the expected objectives of practicing specific knowledge and general skills. Both methodologies allow supervising around 15 works (around 45 students) with high expectancies.

Moreover, the results of the second methodology proof the benefits of working with a more guided manner: the dedicated hours to the work are better distributed both for the students and the teachers; the required objectives are more clear for the students; the evaluation process of the teachers is more objective; the rubrics are a good tool to identify the evaluation criteria; the peer-assessment process helps the students to be aware of the quality of their work and being critic with the classmates' work. It is also feel for the teachers' staff that the students prefer working with more guided methodologies although the originality would be reduced.

Due to the cutbacks in the university budgets, these two methodologies are not done during the nowadays academic year, but they are kept as the backup of successful strategies.

Acknowledgments The authors want to thanks to all teachers who have participated during all these years in Theory of Machines and Mechanisms, making valuable contributions for making possible this supervised work.

References

1. Universitat Politècnica de Catalunya. Departament d'Enginyeria Mecànica. Programa d'Anàlisi de Mecanismes, http://em.upc.edu/docencia/estudis-de-grau/etseib/teoria-de-maquines/programa-danalisi-de-mecanismes.-pam (5th March 2013)
2. Universitat Politècnica de Catalunya. Departament d'Enginyeria Mecànica. Presentació d'informes científics i tècnics, http://em.upc.edu/docencia/estudis-de-grau/etseib/teoria-de-maquines/fascicles-i-normes (5th March 2013)
3. Jordi-Nebot, L. et al.: Enhancements in evaluating small group work in courses with large number of students. Machine theory at industrial engineering degrees. J. Tech. Sci. Edu., **3**(1): 86–96 (2013)
4. Artobolevski, I. I.: Mecanismos en la técnica moderna, Editorial Mir, vol. 1, Moscow, y 2 (1977)

MIX
Papier aus verantwortungsvollen Quellen
Paper from responsible sources
FSC® C105338
FSC
www.fsc.org

If you have any concerns about our products,
you can contact us on
ProductSafety@springernature.com

In case Publisher is established outside the EU,
the EU authorized representative is:
Springer Nature Customer Service Center GmbH
Europaplatz 3, 69115 Heidelberg, Germany

Printed by Libri Plureos GmbH
in Hamburg, Germany